Maths Skills
for AS and A Level
Psychology

Cara Flanagan

OXFORD
UNIVERSITY PRESS

Contents

How to use this book   3

Required mathematical skills
for psychology (Ofqual)   4

Section 1: Arithmetic computation   6
1.1   Fractions   6
1.2   Decimals and decimal places   8
1.3   Percentages   10
1.4   Ratios   12
1.5   Standard form   14
1.6   Estimate results   16
1.7   Order of magnitude calculations   18
1.8   Significant figures   19
1.9   Algebra   21
Summary questions for
section 1   23

Section 2: Handling data   25
2.1   Quantitative and qualitative data   25
2.2   Primary and secondary data   27
2.3   Sampling participants   29
2.4   Sampling observations   31
2.5   Measures of central tendency: mean,
median, and mode   33
2.6   Levels of measurement   35
2.7   Measures of dispersion: range and
standard deviation   37
Halfway summary questions   39
2.8   Frequency tables   41
2.9   Ranking data   42
2.10 Frequency diagrams: bar charts
and histograms   43
2.11 Normal distribution   46

2.12 Skewed distributions   48
2.13 Scatter diagrams   50
Summary questions for
section 2   53

Section 3: Inferential statistics   55
3.1   Simple probability and the
null hypothesis   55
3.2   Hypothesis testing   57
3.3   Error types   59
3.4   Using inferential statistical tests   61
3.5   Sign test   63
3.6   Wilcoxon test   65
3.7   Mann-Whitney test   67
3.8   Related t-test   72
3.9   Unrelated t-test   75
3.10 Chi-squared test   78
3.11 Spearman's test   81
3.12 Pearson's test   85
Summary questions for
section 3   87

References   89

Answers   90

Index   106

The Ofqual table on pages 4–5 is the
same for all psychology specifications,
but the research methods content of the
different specifications varies somewhat,
for example, names of different sampling
methods. There are notes about this in the
appropriate sections.

How to use this book

This workbook has been written to support the development of key mathematics skills required to achieve success in your A Level Psychology course. It has been devised and written by a teacher and the practice questions included reflect the **exam-tested content** for AQA, OCR, Edexcel, and WJEC/Eduqas specifications.

The workbook is structured into sections, with each section relating to a topic in the psychology specification. Then, each topic covers a mathematical skill or skills that you may need to practise. Each topic offers the following features:

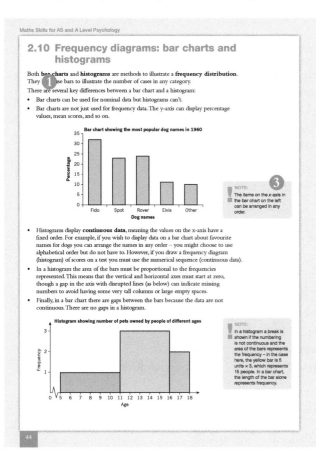

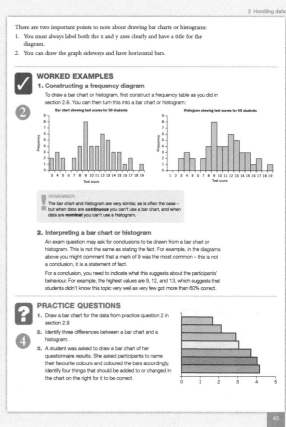

① *Opening paragraph* outlines the mathematical skill or skills covered within the spread.

② *Worked example* – almost all spreads have one or more worked examples. The worked examples will be annotated.

③ *Remember or note* – useful boxes that will offer you tips, hints, and other snippets of useful information.

④ *Practice questions* are provided for all topics, with answers at the end of this book.

Required mathematical skills for psychology (Ofqual)

From 2015, Ofqual (The Office of Qualifications and Examinations Regulation for England) set out specific criteria to be covered in the document *GCE Subject Level Conditions and Requirements for Psychology*.

The requirements for psychology appear on page 38 of that document and are shown in the table below. The contents below have been presented in a different order to the Ofqual document so that similar items are grouped together.

	Mathematical skills	Exemplification of mathematical skill in the context of A level psychology (assessment is not limited to the examples given below)	Where covered in this book
Arithmetic and numerical computation	Recognise and use expressions in decimal and standard form.	For example, converting data in standard form from a results table into decimal form in order to construct a pie chart.	1.2, 1.5
	Use ratios, fractions and percentages.	For example, calculating the percentages of cases that fall into different categories in an observation study.	1.1, 1.3, 1.4
	Estimate results.	For example, commenting on the spread of scores for a set of data, which would require estimating the range.	1.6
Handling data	Use an appropriate number of significant figures.	For example, expressing a correlation coefficient to two or three significant figures.	1.8
	Find arithmetic means.	For example, calculating the means for two conditions using raw data from a class experiment.	2.5
	Construct and interpret frequency tables and diagrams, bar charts and histograms.	For example, selecting and sketching an appropriate form of data display for a given set of data.	2.8, 2.10
	Understand simple probability.	For example, explaining the difference between the 0.05 and 0.01 levels of significance.	3.1
	Understand the principles of sampling as applied to scientific data.	For example, explaining how a random or stratified sample could be obtained from a target population.	2.3, 2.4
	Understand the terms mean, median and mode.	For example, explaining the differences between the mean, median and mode and selecting which measure of central tendency is most appropriate for a given set of data. Calculate standard deviation.	2.5
	Use a scatter diagram to identify a correlation between two variables.	For example, plotting two variables from an investigation on a scatter diagram and identifying the pattern as a positive correlation, a negative correlation or no correlation.	2.13

Handling data	Use a statistical test.	For example, calculating a non-parametric test of differences using the data from a given experiment.	3.4–3.12
	Make order of magnitude calculations.	For example, estimating the mean test score for a large number of participants on the basis of the total overall score.	1.7, 3.8–3.12
	Distinguish between levels of measurement.	For example, stating the level of measurement (nominal, ordinal or interval) that has been used in a study.	2.6
	Know the characteristics of normal and skewed distributions.	For example, being presented with a set of scores from an experiment and being asked to indicate the position of the mean (or median, or mode).	2.11, 2.12
	Select an appropriate statistical test.	For example, selecting a suitable inferential test for a given practical investigation and explaining why the chosen test is appropriate.	3.4
	Use statistical tables to determine significance.	For example, using an extract from statistical tables to say whether or not a given observed value is significant at the 0.05 level of significance for a one-tailed test.	3.4–3.12
	Understand measures of dispersion, including standard deviation and range.	For example, explaining why the standard deviation might be a more useful measure of dispersion for a given set of scores, e.g. where there is an outlying score.	2.7
	Understand the differences between qualitative and quantitative data.	For example, explaining how a given qualitative measure (for example, an interview transcript) might be converted into quantitative data.	2.1
	Understand the difference between primary and secondary data.	For example, stating whether data collected by a researcher dealing directly with participants is primary or secondary data.	2.2
Algebra	Understand and use the symbols: $=, <, <<, >>, >, \propto, \sim$.	For example, expressing the outcome of an inferential test in the conventional form by stating the level of significance at the 0.05 level or 0.01 level by using symbols appropriately.	1.9
	Substitute numerical values into algebraic equations using appropriate units for physical quantities.	For example, inserting the appropriate values from a given set of data into the formula for a statistical test, e.g. inserting the N value (for the number of scores) into the Chi Square formula.	1.9
	Solve simple algebraic equations.	For example, calculating the degrees of freedom for a Chi Square test.	1.9, 3.8–3.12
Graphs	Translate information between graphical, numerical and algebraic forms.	For example, using a set of numerical data (a set of scores) from a record sheet to construct a bar graph.	2.10–2.13
	Plot two variables from experimental or other data.	For example, sketching a scatter diagram using two sets of data from a correlational investigation.	2.10–2.13

Note that these examples are provided by Ofqual – they are
not necessarily the way the skills will be examined

1.1 Fractions

When you divide a whole into parts you get a fraction – a fraction is a part of a whole. If you cut a pie into 6 slices each slice is 1 part of 6 = $\frac{1}{6}$

The 'whole' could also be a whole class or a whole sample of participants. All parts of the whole must be the same if you are expressing them as a fraction.

A fraction consists of two numbers: the **numerator** on the top and the **denominator** on the bottom: $\frac{numerator}{denominator}$

For example, in the fraction $\frac{11}{15}$, 11 is the numerator and 15 is the denominator.

WORKED EXAMPLES

You need to know the following four ways to manipulate fractions.

1. Work out a fraction

If there are 16 female participants and 20 male participants in a study, what fraction of the participants are females?

In this case the 'whole' is the total number of participants = 16 + 20 = 36

The 'part' that is female is 16

So the fraction is 16 out of 36 or $\frac{16}{36}$

2. Simplify the fraction

It is easier to understand a fraction if you 'reduce' it (simplify it).

For example, $\frac{30}{60}$ is easier to understand if you say it is the same as $\frac{1}{2}$

To 'reduce' a fraction, see if there is a number that will divide exactly into both the numerator *and* the denominator.

In the case of the fraction $\frac{16}{36}$, the greatest **common factor** is 4

You can reduce your fraction to $\frac{4}{9}$ by dividing the numerator and the denominator by 4:

$$\frac{16}{4} = 4 \text{ and } \frac{36}{4} = 9$$

 REMEMBER:
A common factor is a number that can be divided into two (or more) different numbers without leaving a remainder.

3. Working backwards

There are five participant groups in a study and each group contains equal numbers of participants. That means that $\frac{1}{5}$ of the total participants are in each group. If the total number of participants is 45, how many are in each group?

To work out $\frac{1}{5}$ of 45, you divide 45 by 5, which equals 9

(Alternatively, you can multiply 45 by the decimal fraction for a fifth, which is 0.2 – see section 1.2.)

So there are 9 participants in each group.

4. Working out more complex fractions

What if you wanted $\frac{3}{5}$ of your 45 participants to be female?

Divide the whole by the denominator = $\frac{45}{5} = 9$

Then multiply 9 by the numerator = $9 \times 3 = 27$

So, 27 participants would be female and the rest would be male ($45 - 27 = 18$).

What if the total group was 46 participants?

Divide the total by the denominator = $\dfrac{46}{5} = 9.2$

Then multiply by the numerator = $9.2 \times 3 = 27.6$

So 27.6 would be females. You can't have 27.6 people, so you need to round the number to the nearest whole number. So the answer would be 28 female participants.

(See section 1.2 on when to round numbers up or down.)

> **NOTE:**
> Rule for rounding up or down: consider the digits to be removed. If these digits are equal to or greater than 5, then round up. Otherwise round down.

PRACTICE QUESTIONS

1. Work out the following:

   a. You have 4 red balls and 16 blue balls. Calculate what fraction of the balls are red.

   b. Simplify your answer to part a.

   c. What is $\dfrac{28}{112}$ in its simplest form?

   d. Calculate $\dfrac{1}{8}$ of 48

   e. Calculate $\dfrac{3}{8}$ of 48

> **REMEMBER:**
> Always show your working out in an answer as there are often marks in an exam for doing the right calculation even if the final answer is wrong.

2. A research project investigated student attitudes towards smoking. A sample of 105 students answered a questionnaire.

   a. 50 students were under 16. Calculate what fraction of the students were under 16. Give your answer in its lowest form (i.e. simplify the fraction).

   b. 80 students said they had smoked at some time in their lives. Calculate what fraction of the students had smoked at some time.

   c. $\dfrac{1}{3}$ of the students said one of their parents smoked. Calculate how many students had parents who smoked.

   d. $\dfrac{2}{5}$ of the sample were female. How many students were female?

3. A researcher wants to collect a sample for a research project by representing groups of people according to their frequency in the population. This is called a **stratified** or **quota sample** (see section 2.3). The researcher collects data from a group of local primary schools:

Age of children	5- to 6-year-olds	7- to 8-year-olds	9- to 10-year-olds	11- to 12-year-olds
Number of children	336	294	252	126

   a. Calculate what fraction of the children are 9- to 10-year-olds. Give the answer in its simplest form.

   b. If the researcher was going to use a sample of 100 children, calculate how many 9- to 10-year-olds they would need if they wanted to use the same fraction as in the initial project.

   c. Calculate what fraction of the children are 11- to 12-year-olds. Give the answer in its simplest form.

   d. Calculate how many 11- to 12-year-olds the researcher would need if they were going to sample 100 children using the same fraction as in the initial project.

1.2 Decimals and decimal places

The word 'decimal' means a system of numbers based on multiples of tens, for example, thousands, hundreds, tens, units, tenths, hundredths, thousandths – these are all multiples of 10

The decimal point is the key feature in decimal numbers. Digits to the left of the decimal point are **whole numbers**, for example, 1s, 10s, 100s. Digits to the right represent **fractions**, for example, $\frac{1}{10}, \frac{1}{100}, \frac{1}{1000}$

The number 0.56 represents $\frac{56}{100}$ because the 6 is in the $\frac{1}{100}$ position.

The number 0.06 represents $\frac{6}{100}$

Zeros are used as place holders and tell you the number of zeros in the denominator. For example, 0.6 is $\frac{6}{10}$, but with an extra zero 0.06 is $\frac{6}{100}$

The term **decimal places** (d.p.) refers to the number of digits to the right of the decimal point. Decimal fractions, such as 0.06, are sometimes just called **decimals**.

WORKED EXAMPLES

You should know these four ways of working with fractions and decimal fractions.

1. Work out the decimal fraction of a whole

What is 0.56 of 85?

0.56 is $\frac{56}{100}$

Multiply the numerator by the whole = 56 × 85 = 4760

Divide this by the denominator = $\frac{4760}{100}$ = 47.6

Alternatively, multiply the decimal fraction by the whole (0.56 × 85) = 47.6

REMEMBER:

To divide by 100 move the decimal point 2 places to the right.

To divide by 1000 move the decimal point 3 places to the right.

To multiply by 1000 move the decimal place 3 places to the left.

2. Work out the fraction equivalent of a decimal

What is 0.56 as a fraction?

0.56 = $\frac{56}{100}$

You can simplify this further by using the greatest common factor of 4. Divide both the numerator and denominator by 4 to get $\frac{14}{25}$

3. Work out the decimal equivalent of a fraction

What is $\frac{3}{5}$ as a decimal?

Divide the numerator by the denominator = $\frac{3}{5}$ = 0.6

Or, a bit more complex:

What is $\frac{6}{11}$ as a decimal?

Divide the numerator by the denominator = $\frac{6}{11}$ = 0.545 4545 45…

If the answer is very long you need to make a decision about the number of decimal places to include. (See worked example 4 on how to round numbers up or down.)

NOTE:

You can work out the decimal equivalent of a fraction using a calculator.

4. Give your answer to a set number of decimal places (d.p.)

In the previous example, $\frac{6}{11}$ was equivalent to 0.545 454 545…

Giving an answer to one or two decimal places is considered precise enough in psychology.

One decimal place means 1 digit to the right of the decimal point, for example, 0.5~~45 4545 45~~, removing the crossed out numbers to get 0.5

Two decimal places means 2 digits to the right of the decimal point, for example, 0.54~~5 4545 45~~. This time the crossed out numbers are greater than halfway to the next number, so you also need to round the answer: 0.55 is closer to 0.545 4545 45 than 0.54 is. Therefore, the answer to two decimal places is represented as 0.55

You must **round up** when the next number you are cutting off is 5 or above, but **round down** when the next number you are cutting off is below 5.

So, for the above example:

- Two decimal places would be 0.55
- Three decimal places would be 0.545
- Four decimal places would be 0.5455

> **REMEMBER:**
> Rule for rounding up or down: consider the digits to be removed. If these digits are equal to or greater than 5, then round up. Otherwise round down.

PRACTICE QUESTIONS

1. Work out the following:

   a. Calculate 0.22 of 44

   b. Write $\frac{3}{4}$ as a decimal.

   c. Write $\frac{5}{6}$ as a decimal. Give the answer to one decimal place.

   d. Write $\frac{6}{7}$ as a decimal. Give the answer to two decimal places.

   e. Write $\frac{3}{13}$ as a decimal. Give the answer to one decimal place.

2. Anita Singh was conducting a questionnaire on happiness as part of her degree dissertation. She found that 77 of her respondents were female and 31 were male.

   a. State the number of female respondents as a decimal fraction of the whole. Give your answer to three decimal places.

   b. State the number of male respondents as a decimal fraction of the whole. Give your answer to three decimal places.

   c. Anita found that the mean score of happiness for all participants was 4.029 Give this score to one decimal place.

3. Give the following numbers to two decimal places:

   a. 12.1409

   b. 4.916

   c. 0.398

   d. 30.003

   e. 26.1056

> **REMEMBER:**
> Your understanding of arithmetic computation can be assessed in various ways, but it is most likely to be in the context of a research study (as in question 2).

1.3 Percentages

'Per cent' means 'out of one hundred'.

The symbol % means 'per cent'.

Therefore 25% means 25 out of 100, or $\dfrac{25}{100}$

This is another way to represent a part of a whole (i.e., a fraction).

WORKED EXAMPLES

You should be able to manipulate percentages, decimal fractions and fractions in the following ways.

1. Change a percentage to decimal fraction

All percentages can be easily converted to a decimal because 25% means 25 out of 100. To turn a percentage into a decimal, divide it by 100 (move the decimal point two places to the left to divide by 100).

For example, $\dfrac{25}{100}$ (or $\dfrac{25.0}{100}$) = 0.25

Similarly, 83% (83.0%) is 0.83 and 49.5% is 0.495

> **NOTE:**
> Take care with decimal fractions.
> 5% is not 0.5
> 5% is 0.05

2. Change a decimal fraction to a percentage

To turn a decimal fraction into a percentage, multiply by 100 (move the decimal point two places to the right) and include the per cent symbol.

For example, 0.34 × 100 = 34%

Similarly, 0.28 is 28% and 1.36 is 136%

3. Change a percentage to a fraction

You can change any percentage to a fraction by removing the % and putting 100 as a denominator.

For example, 25% is $\dfrac{25}{100}$

Remember you can reduce a fraction to its simplest form, for example, $\dfrac{25}{100} = \dfrac{1}{4}$

4. Change a fraction to a percentage

To turn a fraction into a percentage, divide the **numerator** by the **denominator** to get the decimal fraction and then multiply that by 100

For example for the fraction $\dfrac{5}{7}$, divide 5.00 by 7 = 0.71429... and multiply by 100 = 71.43% (2 d.p.)

5. Working out a percentage

For example, if you have 32 flowers and 5 are blue, what percentage are blue?

You know that 5 out of 32 are blue = $\dfrac{5}{32}$

Divide 5 by 32 = 0.15625

Multiply by 100 to get a percentage = 15.625% = 15.6% (1 d.p.)

6. Using a percentage to calculate an answer

If you have 32 flowers and want 15% to be blue, you want to know $\frac{15}{100}$ of 32

$$\frac{15}{100} = 0.15 \times 32 = 4.8$$

As you're not interested in partial flowers, you need to round up the answer to 5

PRACTICE QUESTIONS

1. Work out the following:

   a. You have 4 red balls and 16 blue balls. Calculate what percent are red.

   b. You have 16 balls and want 50% to be red. Calculate how many red balls that would be.

   c. You have 17 balls and want 50% to be red. Calculate how many red balls that would be.

   d. Give 0.158 as a percentage.

   e. Give 0.2 as a percentage.

   f. Give 2% as a decimal fraction.

   g. Give $\frac{1}{8}$ as a percentage.

   h. Give $\frac{3}{8}$ as a percentage.

   i. Give 64% as a fraction.

2. In a psychological study, 23 students are observed in a lesson. During the lesson 10 students are seen to be texting, 18 students answer questions from the teacher, 5 students get out of their seats, and 20 of the students talk to someone else.

   a. Give the fraction of students that are seen to be texting.

   b. Calculate the percentage of students in each of the **four** observation categories. Give each answer to the nearest whole number.

   c. Approximately 80% of the students said they enjoyed the lesson. Calculate how many students that would be.

3. A classic study in psychology (Dement and Kleitman, 1957) was conducted to record how often people reported having dreams during their REM sleep (sleep which occurs when the eyes are rapidly moving). The table below records the results for three participants:

Participant	Dream recalled	No dream recalled	Total number of dreams
DN	17	9	26
IR	26	8	34
KC	36	4	40
Totals	79	21	100

   a. For each of the 3 participants, calculate the percentage of REM sleep during which a dream was recalled. Give all answers to one decimal place.

   b. Outline two conclusions you can draw from these results.

   c. Suggest why it is better to look at the percentages instead of the actual results.

1.4 Ratios

Another way to express a part of a whole (i.e., a fraction) is to use ratios, which are written in the form **a:b**.

- **Part-to-whole** ratios give the part (a) in relation to the whole (b).

  For example, in a group of 40 people there are 30 women. The ratio of women to the whole group is 30:40

- **Part-to-part** ratios give both parts.

  For example, in the same group of 40 people the ratio of women to men will be 30:10

  The sum of the parts in this type of ratio should equal the whole.

As with fractions, ratios can be reduced to their lowest common form. In the examples above you can reduce them:

- Part-to-whole – a ratio of 3:4
- Part-to-part – a ratio of 3:1

The order of the ratio is important, so 3:1 is not the same as 1:3

You must always state what the numbers in a ratio represent.

For example:

- A ratio of 3:1 women to men.
- A ratio of 1:3 men to women – this ratio is the same as the one above.
- A ratio of 1:3 women to men – this ratio is not the same as the other two.

NOTE:
Ratios are often used in sport to express the 'odds' of a certain result. If a racehorse is given 5:1 odds in a race, this means the odds are not in favour of the horse winning. It means if the race was run 6 times you would expect the horse to lose five times and win once.

WORKED EXAMPLES

There are three ways you need to be able to use ratios.

1. Express a part-to-whole relationship

For example, the whole could be a whole cake or a whole experiment.

- In a whole cake there may be 600 g of dry ingredients. If sugar makes up 200 g of the dry ingredients then the ratio of sugar to the whole amount of dry ingredients is 200:600 or 1:3.

- In a whole experiment there may be 75 participants. If 30 of them are under 30-years-old then the ratio of participants under 30 to the total sample is 30:75 or 2:5.

2. Express a part-to-part relationship

- In the example of the cake given above, you may have 200 g sugar, 200 g flour, and 200 g butter. The part-to-part ratio is therefore 200:200:200 or 1:1:1 of sugar, to flour, to butter.

- In the case of the experiment given above, you have 30 people under 30 and 45 people over 30. The part-to-part ratio for this is 30:45 or 2:3 people under 30 to people over 30.

3. Using ratios in calculations

Ratios can be used to calculate the numbers of participants required in a study.

For example, you might want to reflect the composition of a target population in a sample (as in a **stratified** or **quota sample**, see section 2.3). Let's say in the target population there are 3 times as many women as men. This is a ratio of women to men of 3:1.

If the sample size is going to be 32, then you divide 32 by 4 (because the parts are 3 + 1 = 4). The answer is 8, and therefore:

The number of women you want = 8 × 3 = 24

The number of men you want = 8 × 1 = 8

PRACTICE QUESTIONS

1. Work out the following, giving all answers in their simplest form:

   a. Simplify 20:120

   b. You have 4 red balls and 16 blue balls. Give the part-to-part-ratio.

   c. Give the ratio of red balls above as a part-to-whole ratio.

   d. Express the ratio of blue balls above as a part-to-whole ratio.

   e. You have 21 chocolates. You want to divide these between two children in a ratio of 3:4. Calculate how many you would give to each child.

2. In a psychological study there are 80 participants, 50 are girls and 30 are boys.

   a. Give the ratio of girls to boys.

   b. Another researcher aims to repeat the study but this time the sample will be 120 participants. If the ratio from part (a) is maintained, calculate how many girls should take part.

   c. In a further study there will be 39 participants. Calculate how many should be girls to maintain this same ratio.

3. A study was conducted looking at the effect of reading on a child's self-esteem. The study found that 17 children who were proficient readers had high self-esteem whereas 6 of the proficient readers had low self-esteem.

   a. Give the ratio of high to low self-esteem in proficient readers.

   b. Give the ratio of those with high self-esteem to all proficient readers.

   c. Calculate the fraction of proficient readers who have a low self-esteem.

   d. Calculate the percentage of proficient readers who have a low self-esteem. Give the answer to one decimal place.

1.5 Standard form

Standard form is a method of expressing very large or very small numbers by focusing on their magnitude.

When you have a very large number, such as the number of cells in the brain (about 86 billion), you are more interested in the 'billion' than whether there are 86 or 56 billion. 86 billion is a lot more than 86 million, whereas the difference between 86 and 56 billion is less significant.

In addition, if you write 86 000 000 000 000 you would need to count the zeros to work out the number (86 trillion). If you write it in standard form it is easier to see the magnitude.

- You write 86 million as 8.6×10^7
- You write 86 billion as 8.6×10^{10}
- You write 86 trillion as 8.6×10^{13}

Standard form consists of: (A number between 1 and 10) $\times$ ($10^{\text{power of x}}$)

Mantissa (base figure)   **Exponent**

The 'power of x' part (the exponent) shows how many places you need to move the decimal point to the left or the right to produce the mantissa.

You can round the base figure (the mantissa) to one or more decimal places or the nearest whole number. For example, 6 423 204 could be given as 6×10^6 or 6.4×10^6 or 6.42×10^6

The mantissa is always a number between 1 and 10 (6, 6.4, and 6.42 are all between 1 and 10).

> **! NOTE:**
> If you do a large calculation on your calculator you will be given the answer in standard form.
>
> For example, try:
>
> $12\,300 \times 6700$
>
> You should see 8.241e7 or 8.241 7 which means 8.241×10^7
>
> This is the same as 82 410 000

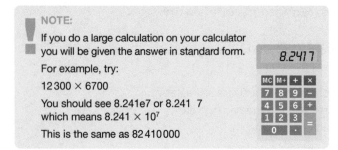

WORKED EXAMPLES

There are three ways you need to be able to use standard form.

1. Changing a very large number to standard form

Consider the number 63 900 000 000

- **Step 1**: Write down the number between 1 and 10 that can be derived from the number to be converted. In this case it would be 6.39 or you could use 6.4 (rounded up) or you could just use 6.

- **Step 2**: Place a red dot after the digit 6 and work out how many times you would have to shift the decimal point to the *left* to get to the red dot – this requires 10 shifts.

$$6.3900000000.$$

- **Step 3**: 63 900 000 000 is therefore written in standard form as: 6.39×10^{10} (or you could use 6.4×10^{10} or 6×10^{10}).

2. Changing standard form back to a very large number

Consider the number 1.9×10^6

To change this back to a number written out in full, you shift the decimal to the right as far as you can go (to reach the red dot), adding zeros as you move 6 times. This results in 1 900 000

$$1.\overset{\frown\frown\frown\frown\frown\frown}{9\,0\,0\,0\,0\,0}.$$

3. Changing a very small number to standard form

Consider the number 0.000 009 3

- **Step 1**: Write down the number between 1 and 10 that can be derived from the number to be converted. In this case it would be 9.3 (or you could use just 9).

- **Step 2**: Place a red dot between the 9 and 3 (0.000 009.3). Now work out how many times you have to shift the decimal point to the *right* to arrive at the red dot. For this number it requires 6 shifts.

$$0.\overset{\frown\frown\frown\frown\frown\frown}{000\,009}.3$$

- **Step 3**: 0.000 009 3 is therefore written in standard form as: 9.3×10^{-6} or 9×10^{-6}

> **NOTE:**
> Notice the sign of the **exponent** is now negative to indicate the decimal point moving to the right rather than moving to the left.

PRACTICE QUESTIONS

1. Give the following numbers in standard form:

   a. 100

   b. 300

   c. 4 000 000

   d. 3600 (give two possible answers)

   e. 6672 (give three possible answers)

   f. 0.000 039 (give two possible answers)

   g. 0.001

2. Give the following numbers in full:

   a. 1×10^6

   b. 5.63×10^4

   c. 2.9×10^9

   d. 1×10^{-5}

   e. 5.63×10^{-4}

   f. 2.9×10^{-9}

1.6 Estimate results

Often you don't need to make exact calculations, and instead a 'ballpark' figure will do. So you make an **estimate** of the results of a calculation. An estimate is not a guess – it is based on rounding figures up or down (see right). This is done so you can calculate the results easily without a calculator.

Estimating results is particularly important as a way of checking a calculation. When making a calculation it is invaluable to have a rough idea of what the results are likely to be so that, if you did make an error in the real calculation, you would notice it.

 REMEMBER:
Rule for rounding up or down: consider the digits to be removed. If these digits are equal to or greater than 5, then round up. Otherwise round down.

 ## WORKED EXAMPLES

Example 1

What is 29% of 351?

Before doing the calculation you can estimate what the answer will be:

29% is roughly $\frac{1}{3}$

351 is roughly 360 (360 is chosen because it is easily divided by 3)

Therefore you would estimate $\frac{360}{3}$ is 120

The precise answer is 101.79

 NOTE:
There are no right answers when you are estimating. However, your estimates need to make sense.

In the example on the left it would not make sense to estimate that 29% is roughly $\frac{1}{2}$ because it is nowhere near that.

It would also not make sense to estimate 351 as roughly 400 when 360 would work just as well in this calculation.

Example 2

What is 3948 × 18?

3948 can be rounded to 4000, and 18 can be rounded to 20

To do this without a calculator you can multiply 2 × 4 (= 8) and add the correct number of zeros from both the rounded numbers (there are 4 zeros). The estimated answer would therefore be 80 000

The precise answer is 67 464

 NOTE:
If you have rounded both figures up your answer will be quite a lot larger than the precise answer.

Example 3

What is 234 522 divided by 43?

234 522 can be rounded to 200 000, and 43 can be rounded to 40

Divide 20 by 4 (= 5), and take the number of zeros not used in the larger number (you divided 20 instead of 200 000 therefore there were 4 'unused' zeros) and subtract the number of zeros not used in the smaller number (1 zero), giving 3 zeros.

You would therefore estimate the answer would be about 5000

The precise answer is 5454

PRACTICE QUESTIONS

1. Tasnem finds arithmetic challenging but she loves making estimations and can't understand why her friend finds it so difficult. Tasnem gives her friend some questions to help.

    a. Tasnem thinks of the number 5648. Suggest whether it is best to use 5000, 5500, or 6000 as an estimate in a calculation using this number. Explain your choice.

    b. Tasnem thinks of the number 462. Suggest what would be the best number to use as an estimate in any calculations – 400, 450, 460, or 500. Explain your choice.

2. For each of the following calculations, (i) estimate the result and then (ii) use a calculator to get the actual result:

    a. 239×162

    b. $16 \times 453\,262$

    c. $\dfrac{239}{8}$

    d. $\dfrac{3}{16}$ of $44\,381$

    e. $\dfrac{5}{8}$ of 267

    f. 78% of 3527

3. Now use a calculator to calculate the actual result for parts a–f in question 2. How close were your estimates? Can you explain why some are not very close?

1.7 Order of magnitude calculations

The word **magnitude** refers to size. Order of magnitude calculations involve making a comparison between the size of two numbers. You use standard form because order of magnitude values are given by the powers of 10 included in the standard form.

The numbers 15 100 and 11 217 are in the same order of magnitude because they are both 10 000s (i.e., they are both 10 to the power of 4 or 10^4).

The numbers 15 100 000 and 11 217 are not the same order of magnitude. The first one is a lot bigger. In fact it is 3 times bigger. You know this because 15 100 000 is approximately 2×10^7 and 11 217 is approximately 1×10^4, and $7 - 4 = 3$. All you are interested in is the difference between the **exponents**.

WORKED EXAMPLES

You need to be able to make order of magnitude comparisons.

1. Comparing big numbers

First, convert the numbers being compared to standard form. For example, to compare 304 538 000 and 41 540:

$304\,538\,000 = 3 \times 10^8$ and $41\,540 = 4 \times 10^4$

Next, subtract the smaller exponent from the larger (i.e. 'to the power of 8' minus 'to the power of 4'):

$8 - 4 = 4$

The larger number is 4 orders of magnitude bigger than the smaller one (i.e. 10 000 times larger).

2. Comparing small numbers

With numbers smaller than zero you still need to subtract, but remember that two minuses make a plus.

For example:

$0.000\,32 = 3.2 \times 10^{-4}$ and $0.000\,000\,047 = 4.7 \times 10^{-8}$

You subtract one exponent from the other (subtracting a -8 gives $+8$):

$-4 - (-8) = 4$

The smaller number is 4 orders of magnitude smaller (i.e. 10 000 times smaller).

PRACTICE QUESTIONS

1. For each of the following, calculate how many orders of magnitude bigger:

   a. 122 000 is than 42 000

   b. 4 500 000 is than 1831

   c. 0.02 is than 0.000 52

   d. 4 500 000 is than 0.02

1.8 Significant figures

One way to round figures up or down is to specify the number of decimal places required (e.g., 'give your answer to two decimal places') or to say 'give your answer to the nearest whole number'.

Another method is to use **significant figures** (s.f.), which means 'the number of digits of importance'. Significant figures are the digits that provide accuracy starting from the first digit that is not zero.

A significant figure is any non-zero digit or any embedded zero or any trailing zero before the decimal point.

> **REMEMBER:**
> The first significant figure in a number is the first digit that is not zero.
> In 2.34 the first significant figure is 2, and in 0.005 34 the first significant figure is 5

Some quick examples:

- The number 0.029 has two significant figures because the first non-zero digit is 2
- The number 0.02905 has four significant figures because the first non-zero digit is 2 and the next zero is embedded.
- The number 503.0 has four significant figures because the first zero is embedded and a second trailing zero is shown.
- The number 530 has three significant figures because the zero is trailing.
- The number 530.029 has six significant figures because the zeros are embedded.

 WORKED EXAMPLES

Here are some example calculations using significant figures. Each time, the number of significant figures required is underlined so that you can decide whether to round up or not by looking at the digits to be removed. In some cases you need to insert a zero to maintain the correct magnitude for the whole number.

Example 1

- 43 204 to two significant figures:

    The two significant figures will be 43 (<u>43</u> 204) and you must replace the '204' with three trailing zeros. So the answer is 43 000

- 43 504 to two significant figures:

    The two significant figures will still be 43 (<u>43</u> 504) and you must replace the '504' but this time you are required to round up. So the answer is 44 000

Example 2

- 0.000 495 to two significant figures:

    The two significant figures will be 49 (0.000 <u>49</u>5) = 0.000 50 because you round up. You record the final zero because otherwise you don't have two significant figures.

- 14.95 to two significant figures:

    The two significant figures will be 14 (<u>14</u>.95) = 15 because you round up. In this case you don't record the 'lost' digits as they are not part of having two significant figures.

- 4.95 to two significant figures:

  The two significant figures will be 4.95 = 5.0 because you round up and now the zero is a trailing zero because otherwise you don't have two significant figures.

Example 3

- 0.060 49 to two significant figures:

  The two significant figures are 60 (0.060 49) = 0.060 because you don't round up but show a trailing zero to provide two significant figures.

- 0.060 49 to three significant figures:

  The three significant figures are 604 (0.060 49) = 0.0605 because you round up.

- 604.9 to one significant figure:

  The one significant figure is 6 (604.9) = 600 because you don't round up but must have 3 digits so that the 6 still represents the 100s to maintain the same magnitude.

Example 4

- 604.8306 to two significant figures:

  The two significant figures are 60 (604.8306) = 600 because you do not round up and need a trailing zero as a place holder.

- 604.8306 to four significant figures:

  The four significant figures are 6048 (604.8306) = 604.8 because you do not round up.

- 604.8306 to six significant figures:

  The six significant figures are 604830 (604.8306) = 604.831 because you do round up.

NOTE:

It is easy to get confused between decimal places and significant figures.

- 14.9 to two decimal places is 14.90
- 604.8306 to two decimal places is 604.83
- 6 048 306 to two decimal places is 6 048 306.00

Example 5

- 6 048 306 to two significant figures:

  The two significant figures are 60 (6 048 306) = 6 000 000 because you do not round up and need the trailing zeros as place holders.

- 6 048 306 to three significant figures:

  The three significant figures are 604 (6 048 306) = 6 050 000 because you do round up and need the trailing zeros.

- 6 048 306 to six significant figures:

  The six significant figures are 604830 (6 048 306) = 6 048 310 because you do round up the final sixth digit and need a trailing zero.

PRACTICE QUESTIONS

1. Work out the following:

- **a.** 104.565 78 to three significant figures
- **b.** 104.565 78 to five significant figures
- **c.** 2432.8976 to two significant figures
- **d.** 2432.8976 to four significant figures
- **e.** 2432.8976 to six significant figures
- **f.** 6.1467 to two significant figures
- **g.** 6.1467 to three significant figures
- **h.** 0.005 137 9 to two significant figures
- **i.** 0.005 137 9 to four significant figures
- **j.** 0.000 468 to two significant figures
- **k.** 0.0702 to two significant figures
- **l.** 0.0702 to three significant figures

1.9 Algebra

Algebra is the study of mathematical symbols and the rules for manipulating these symbols. It enables you to produce an abstract set of rules and apply these to any situation – such as the formula for an inferential test (these tests are covered in section 3).

Mathematical symbols

- Digits: the digits 1, 2, 3, etc. represent quantities.
- Signs: you know the symbols for add, subtract, multiply, and divide ($+ - \times /$). There is also square root and squared ($\sqrt{36}$ means the square root of 36, and 6^2 means 6 squared, i.e., 6×6).
- You need to know these other symbols listed in the psychology specifications:

= and ~	< and <<	> and >>	≤ and ≥	∝
Equal and approximately equal	Less than and much less than	More than and much more than	Less than or equal to, and more than or equal to	Proportional to

Algebraic equations

Algebra uses the mathematical symbols to describe relationships, which are expressed as equations.

This is an equation that tells you how to convert miles (x) to kilometres (y):

$$x = y \times \frac{5}{8}$$

This is an equation to work out the percentage of a quantity, e.g., if you want to know 15% of y apples:

$$x = \frac{15}{100} \times y$$

WORKED EXAMPLE

In psychology you main use of algebraic equations is in the formulae for descriptive and inferential statistics.

For example, the formula for standard deviation (see section 2.7):

$$s = \sqrt{\frac{\Sigma(x - \bar{x})^2}{n - 1}}$$

You can use this formula to work out the standard deviation for any set of known values. You have to calculate or be given values for $\Sigma(x - \bar{x})^2$ and n

If $\Sigma(x - \bar{x})^2 = 42$ and $n = 11$, then:

$$s = \sqrt{\frac{42}{(11 - 1)}}$$
$$= \sqrt{4.2}$$
$$= 2.05 \ (2 \text{ d.p.})$$

NOTE:

There are other examples of substituting values into equations in sections 3.7 (Mann–Whitney test), 3.8 (related t-test), and 3.11 (Spearman's test).

PRACTICE QUESTIONS

1. Calculate 37.5% of 69 using the equation given above. Give your answer to the nearest whole number.

2. Using the formula given above, calculate the standard deviation of a sample where: $\Sigma(x - \bar{x})^2 = 194$ and $n = 15$

3. The formula for one of the values in the Mann–Whitney test (a type of inferential test) is:

$$U_a = n_a n_b + \frac{n_a(n_a + 1)}{2} - \Sigma R_a$$

Calculate the value for U_a if $n_a = 12$ and $n_b = 10$, $\Sigma R_a = 180$

1. Piliavin *et al.* (1969) conducted a study to see whether people on a subway train would be more likely to help a confederate (a knowing participant in the study) who collapsed if the confederate had a walking stick (appeared disabled) or carried an alcoholic drink (appeared drunk). Piliavin *et al.* found that in 62 of the 65 'disabled' trials people helped the confederate and in 19 out of the 38 'drunk' trials people helped the confederate.

  a. Give the ratio of 'drunk' to 'disabled' trials.

  b. Is this ratio closer to 1:2 or 1:3?

  c. Calculate the fraction of people in the train that helped the 'disabled' confederate.

  d. Convert your answer to part c to a percentage. Give the answer to one decimal place.

  e. Calculate the fraction of people in the train that helped the 'drunk' confederate.

  f. Convert your answer to part e to a percentage. Give the answer to one decimal place.

  g. 87% of the 'disabled' confederates were helped within 70 seconds. There were 65 'disabled' trials. Calculate how many of the confederates were helped within 70 seconds.

  h. There were 103 trials in total. Estimate what percentage of the trials were 'drunk' trials.

2. A human has approximately 86 000 000 000 neurons in their whole nervous system. An African elephant is estimated to have 257 000 000 000 neurons, whereas a lion has an estimated 4 667 000 000 neurons.

  a. Give all three estimates to two significant figures.

  b. Give all three estimates in standard form.

  c. Calculate how many times bigger the number of neurons in an African elephant is compared to a human.

  d. Make the same comparison between a human and a lion.

  e. The cranial capacity of modern humans is 1.4×10^3 cubic centimetres. Write this number in its full form.

3. Loftus and Palmer (1974) showed a film of a car accident to participants and asked them to estimate the speed the cars were travelling when they 'hit' each other. The critical word 'hit' was varied for different participants and was sometimes 'smashed' or 'contacted'.

  a. They found that the mean speed estimate when using the word 'smashed' was 40.8 mph. Give the number of decimal places in this result.

  b. They found that the mean speed estimate when using the word 'hit' was 34.0 mph and for 'contacted' was 31.8 mph. Give the three mean speed estimate results to two significant figures.

  c. There were 45 participants in the study spread between 5 different conditions. Calculate the percentage of participants in each condition.

4. Milgram (1963) investigated obedience to destructive orders. He asked 40 men to deliver increasingly strong electric shocks to a 'learner' who was required to recall word pairs.

  a. 65% of participants delivered the highest level of shock to a 'learner'. Calculate the number of participants who delivered the highest level of shock.

  b. Five participants stopped at 300 volts. Give the percentage of participants who stopped at this level.

  c. Give the ratio of participants who stopped at 300 volts to those who went to the highest level.

d. The answer to part c is a rather awkward ratio. Which ratio is it closest to 1:5 or 1:4?

e. A replication of the study involved 86 participants. Give the number of participants you would expect to be fully obedient if the rate was 65%.

5. Some students carried out a project on memory comparing participants' performance in the morning and afternoon. They gave participants 50 words to memorise and recall after 30 minutes. They found the mean recall in the morning was 32.4082 words and in the afternoon was 20.9901 words.

a. Give both results to two decimal places.

b. Give both results to three significant figures.

c. Express each result as a percentage. Give the answers to the nearest whole number.

d. Estimate what the answers to part c should have been. Show your working.

e. The students repeated the study but this time found that the morning group remembered 48% of the words on average. Calculate how many words that would be.

6. Bandura *et al*. (1961) tested how children responded when watching a model behave aggressively to a Bobo doll. There were 72 children in the study and they were divided equally into 3 groups: a control group, one group that watched a model behave aggressively, and one group that watched a model not behave aggressively.

a. Calculate the fraction of the sample in each of the 3 groups.

b. Calculate how many children were in each of the 3 groups.

c. In the non-aggressive model group and the control group 70% of the participants had zero scores for imitative aggression. Calculate how many participants that is.

7. Raine *et al*. (1997) compared the brains of murderers and non-murderers.

a. 39 of the murderers were men and 2 were women. Estimate what percentage were women.

b. Calculate the exact answer to part a. Give the answer to one decimal place.

c. The part-to-part ratio of murderers to non-murderers was 1:1. Calculate how many non-murderers were in the study.

d. They found that 23 of the murderers had no history of brain injury. Calculate this as a percentage of all the murderers in the study. Give the answer to one decimal place.

8. Some facts about the human body:

a. The ratio of the height of your head to your whole body is about 1:8. For a person who is 168 cm tall, calculate what the height of their head should be.

b. The cell body of a neuron is 0.01 cm long. The axon of a motor neuron is 10000 times longer. Calculate the length of an axon.

c. A synapse is about 30 nanometres wide and a nanometre equals 1×10^{-7} cm. Write 1×10^{-7} out in full, multiply this by 30, and write the answer in standard form.

9. A psychologist was told that the formula for calculating degrees of freedom (*df*) in a chi-squared test is $(r-1) \times (c-1)$, where r is the number of rows in a contingency table and c is the number of columns.

a. How many degrees of freedom are there in a contingency table which has three rows and four columns?

b. How many rows are there in a contingency table if there are six degrees of freedom and two columns?

c. How many columns are there in a contingency table if there are eight degrees of freedom and three rows?

2 HANDLING DATA

2.1 Quantitative and qualitative data

Quantitative data

Quantitative data are numerical information that represents quantities of something, for example, how much, how long, or how many there are of something. Examples of quantitative data are your age, how many hours you work in a week, how highly you rate different TV programmes, and so on.

Strengths	Limitations
+ Easier to analyse quantitative than qualitative data because it can be summarised using, for example, graphs and averages. This tends to make it easier to draw conclusions, such as seeing differences at a glance on a bar chart, or that the mean rating for a particular film was 7 out of 10 + Quantities are a more objective approach because measurements (e.g., height or test score) should be the same no matter who is doing the measuring. This gives the measurements greater validity because they are not biased by the person doing the measurements.	− Quantitative scores may not express views precisely because participants are forced into selecting a number, for example, rating attitudes about capital punishment on a scale of 1 to 10. Therefore, the data collected may have less validity than asking someone to describe their views. − Oversimplifies reality and human experience because it suggests that there are simple answers, i.e., it is a reductionist approach reducing human experience to quantities.

Qualitative data

Qualitative data are information that expresses the 'quality' of things. This includes descriptions, words, meanings, pictures, and so on. Qualitative data can't be directly counted or quantified, though they can be turned into quantitative data by placing the data in categories and counting the frequency.

Strengths	Limitations
+ Represents true complexity of human behaviour because thoughts/behaviours are not reduced to numbers. Instead qualitative methods are used to analyse the data, such as thematic analysis where themes are identified. Such themes could be identified either in a person's conversation, in a film, in a painting, etc. The themes can be counted (thus turning qualitative data into quantitative data). + Provides rich details of how people think and behave because they are given a free reign to express themselves. Therefore, qualitative data can have greater validity.	− More difficult to detect patterns and draw conclusions because of the large variety of information collected, and because words cannot easily be reduced to a few simple points. − Interpreting what people mean is likely to be subjective, because people may not clearly express what they mean or the observer puts their own interpretation on the answers. This lowers the objectivity of the final data collected, reducing validity.

WORKED EXAMPLES

You should be able to identify qualitative and quantitative data.

Example 1

In a questionnaire, the answer to be selected is 'yes' or 'no'. Would the data produced be quantitative or qualitative?

The data produced would be qualitative because it is 'yes' or 'no'. However, the researcher would turn this into quantitative data by counting the number of yes and no answers. The same applies if a fixed list of answers is provided for a question – the answers are qualitative but turned into quantitative data.

Example 2

Do research studies only produce one kind of data?

No, most studies produce a mixture of quantitative and qualitative data. For example Milgram's (1963) study measured obedience by looking at the shock level that participants were prepared to deliver to another person. Thus each participant had a numerical score going up to 300 volts or 450 volts.

Milgram (1963) also reported the things that the participants said, such as 'He's banging in there. I'm gonna chicken out. I'd like to continue but I can't do that to a man'. This is qualitative data that gives insights into the experience of being obedient to a destructive order.

PRACTICE QUESTIONS

1. In each of the following, identify whether the data are quantitative or qualitative:

   a. The total score for the first participant was 20

   b. The participant gave the answer '10'.

   c. The participant gave the answer 'yes'.

   d. Descriptions of the drawings produced by children.

   e. Ratings of the drawings produced by children in terms of artistic flair.

2. A psychologist is designing a questionnaire about eating habits in young children. She plans to ask mothers about their children's attitudes towards food as well as what they like to eat.

   a. Write a question for this questionnaire that would produce quantitative data.

   b. Write a question for this questionnaire that would produce qualitative data.

3. In an observational study the anxiety levels of people in a doctor's waiting room are assessed.

   a. Give an example of quantitative data that might be collected.

   b. Give an example of qualitative data that might be collected.

   c. Explain the strengths of using quantitative data rather than qualitative data in this study.

4. Bandura *et al.* (1961) conducted research on children's imitation of the aggression displayed by a model towards a Bobo doll.

   a. The children were observed playing with toys and the number of their aggressive behaviours was counted. Identify the type of data produced.

   b. The type of aggression displayed was categorised as imitative, partially imitative, non-imitative, or non-aggressive. Bandura *et al.* found that the children who watched the aggressive model were more likely to show imitative aggression than the children who watched a non-aggressive model. Explain in what way the data might be both qualitative and quantitative.

   c. The children who watched the aggressive model hit the Bobo doll had a mean imitative aggression score of about 11, whereas the children who watched a non-aggressive model had a mean score of about 1. Identify the type of data produced.

2.2 Primary and secondary data

Data can be classified as quantitative or qualitative (see section 2.1) but can also be classified as primary or secondary.

Primary data

Primary data are information collected by a researcher specifically for the purpose of the current study. Note that it isn't simply data collected by the researcher but data collected to address the aims of the specific study.

Strengths	Limitations
+ Primary data suits the aims of the study. As researchers know the type of data they need for their investigation, they can identify exactly the kind of information required. + Primary data are authentic as it comes directly from the participants themselves. This means the data may ultimately be more useful for drawing valid conclusions in relation to the study's aims.	− Collecting primary data takes more time and effort compared with using secondary data. Designing and piloting data collection methods and then collecting data are time-consuming and therefore costly. − A researcher may spend time collecting primary data only to discover that the methods used were flawed. It might have been a better use of time and money to adjust the research aims and use secondary data.

Secondary data

Secondary data are information used in a research study that was collected previously for a different study or another purpose (such as using government statistics to study crime). The researcher collects no new data. Effectively such data are 'second-hand'.

Strengths	Limitations
+ Using secondary data saves time in designing and validating data collection methods. If someone has carried out a very similar study before, this means the data collection methods have already been piloted and checked for validity and reliability. Therefore the current research project requires less time and expense. + Using secondary data can allow access to large data sets, for example, records kept by psychiatric hospitals on diagnosis rates of mental disorders. Such data could be used in a study of genetic factors in mental disorder. + There are times when secondary data are the only kind of data to be used – as in a meta-analysis where a researcher is combining data from a large number of previous studies which shared the same aims and methods. By combining data researchers increase their sample size and thus the validity of their conclusions.	− Previously collected secondary data may not quite fit the needs of the current study, may be out-of-date, not quite complete, or of poor quality. Therefore the ultimate results of the current study can lack validity or usefulness. − Using secondary data may waste valuable time because, if it turns out not to be directly relevant to the aims of the current study, that time could have been better spent designing a primary data investigation. − Using secondary data for a meta-analysis can be problematic because the methods used by the different studies may not be sufficiently similar and thus the final conclusions of the meta-analysis can lack validity.

WORKED EXAMPLE

What kind of data is used in a meta-analysis?

Meta-analyses deal with secondary data – the data being analysed has been taken from other previous studies.

For example, van IJzendoorn and Kroonenberg (1988) studied attachment between mothers and children by combining data from 32 studies conducted in 8 different countries. Each of the 32 studies had individually collected primary data using the same method to measure attachment (the Strange Situation). The primary researchers in these studies did not design the data collection method but it is still primary data because they worked out the design of each study (e.g., sampling, training investigators, etc.). Van IJzendoorn and Kroonenberg used this primary data - but, for their study, it became secondary data.

PRACTICE QUESTIONS

1. Identify whether the following studies are using primary or secondary data:

    a. A study on crime using government statistics.

    b. Research on crime which interviews both victims and criminals.

    c. An analysis of the brains of patients with schizophrenia using brain scans taken by doctors when admitting the patients to a psychiatric institution.

    d. An observational study of different animals in a local zoo.

    e. An investigation of animal behaviour using nature films made for the BBC.

2. Gottesman *et al.* (2010) aimed to find out the likelihood of a person developing schizophrenia if one or both parents had been diagnosed with the disorder. They used data from the Danish civil register of all people born in Denmark between 1968 and 1997, and data from the Danish psychiatric register of all psychiatric admissions between 1970 and 2007. The researchers found a 27.3% chance of a person developing schizophrenia if they had two parents with the disorder and a 7% chance if only one parent had the disorder.

> REMEMBER:
>
> When giving the strengths or limitations of any study, don't just apply the strengths/limitations on the previous page. Use these to help you think about what might be good or bad and to develop a specific criticism.
>
> Generic criticisms that can be applied with little thought rarely gain marks.

    a. Explain why the data in this study is classed as secondary data.

    b. Explain one strength of using secondary data for this study.

    c. The number of people with schizophrenia was based on psychiatric admissions. Explain why this might be a limitation of the data.

3. Bahrick *et al.* (1975) investigated the duration of long-term memory by using high school yearbooks. People between 17 and 74 were shown photographs (not all from their school year) and asked to name any of the faces they remembered. People who had graduated 15 years ago were 90% accurate but this declined to about 70% accuracy after 48 years.

    a. Explain why this is an example of primary data.

    b. Explain one strength of using primary data for this study.

2.3 Sampling participants

Sampling refers to taking a smaller number of units from a larger **population** of interest. The aim is to select a sample that is as representative as possible of the total population.

In psychological research, there are two forms of sampling that are used:

1. Selecting a group of participants from a target population (the population of interest).
2. Selecting a set of observations from a larger set of behaviours (the population of interest).

In this section you will look at the first form of sampling – the methods psychologists use to collect a sample of participants. In section 2.4. you will look at the second form of sampling – methods to sample observations.

Opportunity sample

A sample of participants produced by selecting individuals who are most easily available at the time of the study. For example, asking people walking by in the street or in the common room at school.

Strengths	Limitations
+ The most convenient technique to use because it takes little preparation. The first participants available are used. + It may be the only technique possible because the whole target population cannot be listed (as needed for other sampling techniques).	− Inevitably biased because the sample is drawn from a specific part of the target population which has unique characteristics (e.g., shoppers in a high street), and therefore is unlikely to be representative (e.g., people who shop online or people with very little money). − Participants may refuse to take part so the final sample also has the same limitations of a self-selected/volunteer sample.

Random sample

A sample of participants produced using a random sampling technique that ensures every member of the target population has an equal chance of being selected. For example, by giving everyone in the target population a number, putting the numbers in a hat, and drawing out the required number of participants, or by using a random number generator.

Strengths	Limitations
+ Unbiased as all members of the target population have an equal chance of being selected. + It is possible to choose a specific subgroup in the target population first, which makes it easier to list everyone and then select participants.	− Takes more time than some techniques because ideally a list of all members of the target population is needed, and then the participants chosen must be contacted. However, subgroups may be used, as described on the left. − Often not actually random because not all identified participants can be accessed or agree to take part.

Self-selecting (volunteer) sample

A sample of participants produced by asking for people willing to take part. For example, advertising in a newspaper or on a noticeboard for participants.

Strengths	Limitations
+ Convenient way to find willing participants because they are less likely to drop out as they volunteered. + May be a good way to get a specialised group of participants, for example, if studying the behaviour of medical students then put an advert on the noticeboard of a medical school.	− Sample is biased because volunteer participants are likely to be more highly motivated and/or with extra time on their hands than the population in general (called volunteer bias). − Volunteers may also be more willing to be helpful and thus may be more likely to guess the aims of the study and respond to demand characteristics.

Snowball sampling

Current participants recruit further participants from amongst people they know.

Strengths	Limitations
+ Researcher can locate groups of people who are difficult to access, for example, medical students or drug addicts (once you have one person you can then ask them to recommend a friend). + This method enables research to be conducted on unusual behaviours because sufficient participants can be contacted.	− The sample is not likely to be representative of the population because it is based on acquaintances of current participants. Thus it may be a biased sample. − Many of the people contacted may decline to take part which means the method overall is unsuccessful, or you end up with a self-selected (volunteer) sample with all its related limitations.

Stratified and quota sampling

A sample of participants selected from different subgroups (strata) in the target population in proportion to the subgroup's frequency in that population. Subgroups, such as age groups, are identified and the number of people in each subgroup in the target population is identified. This is represented as a percentage total of the whole population. For example, 30% of the target population might be 10- to 12-years-old. Then 30% of participants in the study should be in that age group.

Stratified sample – the researcher uses random selection to identify the right number of 10- to 12-year-olds.

Quota sample – selection is done using a non-random technique such as opportunity sampling.

Strengths	Limitations
+ This is the most representative of all the sampling techniques because all subgroups are represented and are in proportion to their numbers in the target population. + Specific subgroups can be chosen according to the variables and aims considered important by the researcher. This increases control over possible extraneous/confounding variables.	− The decision about which subgroups to use may be biased, thus reducing the representativeness of the sample. − It is a very lengthy process and those participants selected may not always agree to take part. This means that it is not used much in psychological research. It is used more commonly in market research or when conducting opinion polls.

Systematic sampling

A sample of participants selected using a fixed, periodic interval that is predetermined. For example, selecting every 6th, 14th, 20th person in a phonebook. The numerical interval (every nth) is applied consistently.

Strengths	Limitations
+ Unbiased as participants are selected using an objective system. + As with random sampling, it is possible to choose a specific subgroup in the target population first. This makes it easier to randomly select participants.	− Not truly unbiased/random unless you first select a number using a random method and start with this individual, and then select every nth individual. − As with random sampling, takes more time than some techniques because you need a list of all members of the target population, and some people may refuse to take part which produces a kind of self-selected (volunteer) sample.

WORKED EXAMPLE

In his research into obedience to Stanley Milgram placed an advertisement in a local newspaper asking people to send him their details if they would be willing to participate in a study on memory and learning.

- What form of sampling did Milgram use?
  Since Milgram was asking for people who were willing to take part in his study, he used a self-selecting (volunteer) sample.

- Why might Milgram's sample have been biased?
  It might have been biased because volunteer participants might be psychologically different from people who do not volunteer, for example they might be more highly motivated and more likely to respond to demand characteristics.

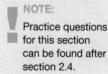
NOTE:
Practice questions for this section can be found after section 2.4.

2.4 Sampling observations

In an observational study, a researcher first of all needs to select a sample of participants using one of the methods described in section 2.3. Then the researcher decides on how to sample the behaviours of each participant. There are two ways of sampling a set of observations from a larger set of behaviours.

Event sampling

This involves counting each time a target behaviour is observed. For example, drawing up a list of behavioural categories and counting (tallying) every time any of the behaviours occurs during a specified time period (e.g., watching for an hour). One target individual may be observed or several observed simultaneously.

Strengths	Limitations
+ Makes observing behaviour more manageable by just noting the occurrence of behaviours of interest. + Event sampling is useful when the behaviour to be recorded only happens occasionally. This means the researcher does not miss events as this would reduce validity.	− Observations may not be representative if the list of events is not comprehensive. This would reduce validity. − If the target events are frequent it may be difficult to record them accurately because many occur simultaneously.

Time sampling

This involves making observations at set time intervals. For example, drawing up a list of behavioural categories and recording observations at fixed intervals like every 5 seconds or every 2 minutes, noting what is happening. Alternatively, the researcher may take a sample at different times of day or month. One target individual may be observed or several observed simultaneously.

Strengths	Limitations
+ Makes observing behaviour more manageable by just noting the occurrence of behaviours at regular intervals rather than watching all the time. + Takes less time than trying to keep a running record especially when target behaviours happen frequently.	− Time sampling may decrease validity because some behaviours are missed if an important behaviour occurs outside of the observation interval. − May not work well if it is difficult to tell when a behaviour has started or ended and therefore the observer is unsure whether to count it.

WORKED EXAMPLES

A psychologist plans to study the work habits of students in the college library.

- How might the psychologist select a sample of students?

    The answer could be any of the methods in section 2.3, However, opportunity sampling is most likely because it would make the most sense to use students who are there.

- How might the psychologist select behaviours?

    One possibility (event sampling) would be to have a list of possible behaviours (staring into space, reading a book, etc.) and tallying each one as they occur, watching one student for 2 hours, and then another student for 2 hours and so on.

    The other possibility (time sampling) is again to use a list of possible behaviours and every 5 minutes tick which one(s) the student is engaged in and do this for 2 hours.

PRACTICE QUESTIONS

These questions are for you to practise what you have learned in sections 2.3 and 2.4

1. Identify the sampling method for selecting participants in each example:

   a. Participants answer an advert in a newspaper.

   b. List all participants in the target population and select every 10th name.

   c. Ask anyone you know if they will take part in your experiment.

   d. Participants selected in different age groups in proportion to their frequency in the population. Selection is done randomly.

   e. Draw up a list of all participants in the target population and put their names in a 'drum'. Select 20 names from this.

   f. A group of psychology students interview shoppers in a shopping centre.

   g. A class of psychology students conduct a study on memory. They put a notice on the noticeboard in the sixth form common room asking for participants who have an hour to spare.

   h. A researcher studies IQ in primary school children by selecting the first 5 names in each class register for every school he visits.

2. For each of the following observations state which sampling procedure (time or event sampling) would be most appropriate and explain why that method would be the best one to use in this particular context.

   a. Observing the different activities young children engage in at a nursery school.

   b. Observing students in a classroom to find out what they spend most of their time doing.

3. A study on memory compared recall in the morning and the afternoon.

   a. Explain how a researcher would use opportunity sampling in this study.

   b. Explain the benefit of using self-selecting/volunteer sampling instead of opportunity sampling.

   c. Explain why stratified sampling would be a better sampling method than opportunity or self-selecting/volunteer sampling.

4. An observational study was conducted looking at the behaviour of pets and their owners.

   a. Explain how a suitable group of pets could be selected to take part in the study.

   b. To observe the behaviour of each pet when the owner was or wasn't in the room, the researcher drew up a list of behaviours, such as sleeping, grooming, playing. Describe how the researcher might use event sampling to record observations.

   c. Explain why time sampling might have been a better way to record the observations than using event sampling.

2.5 Measures of central tendency: mean, median, and mode

The end result of any research is data. You call the data that has been collected 'raw data' until it has been processed in some way. Quantitative data can be processed using **descriptive statistics**:

- **Measures of central tendency** tell you about typical or average values of a data set – see the descriptions of mean, median, and mode below.

- **Measures of dispersion** tell you about the spread of a data set – see section 2.7.

- **Tables and graphs** allow you to see patterns in the data at a glance – see sections 2.8–2.13.

> **NOTE:**
> You need to know the three ways of finding an average (measure of tendency):
> - **Mean:** Adding up all the values and dividing by the total number of values.
> - **Median:** Placing all values in order and selecting the middle value. If there are two middle values, you then need to calculate the mean of these two values.
> - **Mode (modal group):** Identifying the group or groups which is/are most frequent or common.

Measure of central tendency	Strengths	Limitations
Mean	+ A 'sensitive' measure because it reflects the values of all the data in the final calculation. Suitable for **interval data**, see section 2.6.	− Can be unrepresentative of the data set if there are extreme values (outliers).
Median	+ Not affected by extreme scores, as only takes the middle value(s). Suitable for **ordinal data**, see section 2.6.	− Not as 'sensitive' as the mean because not all values are reflected in the final calculation.
Mode	+ Useful when the data are in categories (**nominal data** – see section 2.6), e.g., asking people to state their favourite colour.	− Not a useful way to describe data when there are several modes. For example, if 12 people choose yellow and 12 people choose red.

WORKED EXAMPLES

You need to be familiar with how to calculate mean, median, and mode measures of central tendency as well as when each is the most appropriate measure to use.

1. The mean

The mean is calculated by adding up all the values and dividing this by the total number of values.

For example:

Set A: 5, 8, 8, 11, 12, 16, 17 − the mean = 11.0

Set B: 5, 7, 10, 11, 14, 17, 41 − the mean = 15.0

The second example shows how one extreme value (outlier) can have a large effect on the mean.

2. The median

The median is the middle value of a data set.

For example:

Set A: 5, 8, 8, 11, 12, 16, 17

Set B: 5, 7, 10, 11, 14, 17, 41

There are 7 values in both data sets, therefore the mid value is the 4th one (=11). The median is not affected by the extreme value (outlier).

Sometimes there are two mid values, for example, 5, 7, 8, **11, 12**, 16, 17, 20

In which case you need to then calculate the mean of the two values (11 and 12) = 11.5

3. The mode

The mode is the value that has the highest frequency in the data set.

For example:

Set A: 5, 8, 8, 11, 12, 16, 17

Set B: 5, 7, 10, 11, 14, 17, 41

Set C: 5, 8, 8, 11, 12, 16, 16

Set A has a mode of 8, set B has no mode, and set C has two modes of 8 and 16.

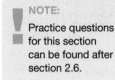

NOTE:
Practice questions for this section can be found after section 2.6.

PRACTICE QUESTIONS

1. Consider the following data: 6, 18, 4, 7, 12, 9, 10, 12, 2, 15, 19

   a. Estimate the mean.

   b. Calculate the mean. Give your answer to two decimal places.

   c. Work out the median.

   d. Work out the mode.

   e. State which measure of central tendency is best to use to reflect the data and explain your decision.

2. Answer the same questions (parts a–e above) for the following data:

   4, 25, 13, 41, 68, 54, 34, 48, 31, 115

2.6 Levels of measurement

The way variables are measured varies in terms of the amount of detail that can be expressed – and this affects the kind of descriptive statistics that can be used.

Nominal Data are in separate categories.	For example, grouping people in your class according to their height – tall, medium, and short people. Within each category you can count how many people there are (= frequency data).
Ordinal Data are ordered in some way.	For example, your class lines up in order of height. The 'difference' between each person is not necessarily the same but you can label them 1st, 2nd, etc. You thereby rank them from highest to lowest. Data from some psychological tests (e.g., IQ tests) are considered to be ordinal because such tests use standardised scores which are not a simple reflection of the number of actual items that were scored correctly. Rating scales are regarded as ordinal because the 'difference' between each item is not the same, i.e., an individual may like the first item a lot more than the second, but there might only be a small difference between the items ranked second and third.
Interval Data are measured using units of equal intervals, but no absolute zero.	For example, the difference between 20 degrees celsius and 21 celsius is the same as the difference between 24 degrees celsius and 25 degrees celsius, that is, 1 degree celsius. Therefore, this temperature scale is equal interval. However, 0 degrees celsius does not mean there is no temperature. Rather, it is an arbitrary zero point, and is the point at which water freezes.
Ratio Equal intervals and an absolute zero.	For example, measuring someone's height in centimetres. A person who is 60 centimetres is twice as tall as a person who is 30 centimetres. Additionally, 0 centimetres is no height, and absolute zero.

WORKED EXAMPLES

Example 1

Edgar's research study looked at whether memory and thinking (i.e. cognitive tasks) are better in the morning or afternoon. He tested one group of participants in the morning and a different group in the afternoon. The test items involved recall of some logical problems.

- What level of measurement is the score on this cognitive test?

  A student who scores, for example, 10 on this test clearly knows more than a student who scores 5 on the test. However, it is difficult to say whether each mark (unit) has an equal interval. Therefore, the scores are probably not at the interval level, but are certainly at the ordinal level.

  If the test had simply involved number of words recalled from a previously viewed list, the data could be said to interval if it could be argued that each word in a list was equal to every other word in the list in terms of memorability. If some words in the list are more memorable than others, and usually they would be, then it might be better to argue that the data again should be treated as ordinal.

Example 2

Chin is a psychology teacher and each year likes to analyse his students' A level results. One analysis is to consider whether his students did worse than their predicted grades.

- What is the level of measurement of the data to be analysed?

  Chin is analysing students' grades (A, B, D, E, F and U). This is data in categories, i.e. nominal.

Example 3

Bettina conducted an experiment on student attitudes to their A level studies. She aimed to compare the differences between boys' and girls' attitudes. Each student was asked to rate how much they enjoyed the psychology course on a scale of 1-5 where 5 represented 'extremely positive'.

- Is the data classed as ordinal or interval?

  It would be difficult to argue that the difference between a rating of 1 and 2 is always the same for each participant as they are making a subjective judgement.

  It would also be impossible to assume that two people who give the same rating have exactly the same level of opinion as each other e.g. Participant A who gives a rating of '5' might be much more in favour of the attitude object than participant B who also gives a rating of '5'.

  Therefore this data is classed as ordinal.

NOTE:
In an exam question, when identifying the level of measurement for a data set, always explain your choice as well as identifying the answer. Sometimes there is no single correct answer and it is your justification that will determine your mark.

Example 4

Although Karim revised hard for his psychology test, he got every question wrong. 'You know nothing about psychology' said his teacher.

- Why is Karim's teacher wrong to say that he 'knows nothing about psychology'?

  A score of 0 on a test is not an 'absolute zero', in the same way that a score of 0 on a personality test does not mean 'no personality'. Karim knows something about psychology, it's just that the questions did not allow him to show that knowledge.

PRACTICE QUESTIONS

1. Give an example of a set of nominal data and explain how the mode would be calculated.

2. Explain why an IQ test score might be classified as ordinal data.

3. Explain the difference between ordinal and interval level data.

2.7 Measures of dispersion: range and standard deviation

A data set can be described by giving the average value (mean, median, or mode), but using the spread of the data as well provides a fuller picture. Neither of the methods below are suitable for nominal data.

Range

The **range** is the difference between the highest and lowest value. This involves arranging the data in order from highest to lowest (or vice versa) and subtracting the lowest number from the highest number.

Strengths	Limitations
+ A convenient way to express how dispersed a data set is because the highest and lowest value are used. + Easy to calculate.	− Affected by extreme values. − Fails to take account of the distribution of the data set. For example, it doesn't indicate whether most numbers are closely grouped around the mean or spread out evenly.

NOTE:

When calculating the range, it is recommended that 1 is added to accommodate the fact that data are often imprecise (the +1 rule).

This is to provide a more precise value for the range when measurements can take fractional values.

Standard deviation (SD)

Standard deviation is also abbreviated to s (standard deviation of the sample) or the Greek letter sigma σ (standard deviation of the population). This quantifies how much each data value differs from the mean. The larger the SD the greater the dispersion.

- **Step 1:** The mean ($\bar{x}$) is calculated for the data set.
- **Step 2:** Take each value in the data set and subtract the mean ($x - \bar{x}$)
- **Step 3:** This value is squared ($x - \bar{x})^2$
- **Step 4:** All squared values are added up $\Sigma(x - \bar{x})^2$
- **Step 5:** The total is divided by $n-1$ where n = the number of values in the data set (the result at this stage is called the **variance**).
- **Step 6:** Finally, the square root is calculated so that the answer is expressed in the same units as the original data.

The formula in full is: $s = \sqrt{\dfrac{\Sigma(x - \bar{x})^2}{n - 1}}$

(See section 1.9 for how to solve formulae.)

REMEMBER:

- Edexcel provide the SD formula on the exam paper, so students may be required to use this.
- Eduqas/WJEC use a different formula – dividing by n instead of n-1. This is a common variation which calculates the standard deviation of the population rather than the sample.
- AQA and OCR students do not need to memorise this formula, but may be asked to substitute values into the formula. The formula would be provided.

Strengths	Limitations
+ A precise measure of dispersion because all the data values are taken into account in the final calculation. + The standard deviation is not difficult to work out if you are using a calculator.	− May hide some of the characteristics of the data set (e.g., extreme values). − Cannot be immediately worked out from the data, whereas the range is quick to identify.

WORKED EXAMPLES

You need to understand the following measures of dispersion, and may be required to calculate them and/or interpret them.

1. Range

In each of the examples below, note how the range helps to *describe* the data, and how the +1 rule is used. (The standard deviation is also given for comparison).

Set A: 5, 8, 8, 11, 12, 16, 17   Mean = 11   Range = 17 − 5 + 1 = 13 (s = 4.4)

Set B: 5, 8, 8, 10, 12, 16, 41   Mean = 14.3   Range = 41 − 5 + 1 = 37 (s = 12.3)

Set C: 1, 1, 2, 5, 8, 12, 41   Mean = 10   Range = 41 − 1 + 1 = 41 (s = 14.3)

2. Standard deviation

All exam boards may require you to substitute values into a formula and do the calculation.

For example, a data set of 15 values produces a value for $\Sigma(x - \bar{x})^2$ of 146. Using the formula given on the previous page (using $n - 1$), the standard deviation would be calculated like this.

$$s = \sqrt{\frac{146}{(15 - 1)}} = 3.2 \text{ (1 d.p.)}$$

> **REMEMBER:**
> You are advised to take a calculator into the exam, and might need it for a calculation like this.

3. Interpreting the standard deviation

Two classes take a memory test. Each class has a different teacher. Their scores are shown below:

Class A: 2, 8, 10, 11, 13, 14, 15, 17, 18 s = 4.9

Class B: 2, 4, 5, 8, 10, 11, 17, 18, 19 s = 6.3

The standard deviation suggests a much wider variation in class B – so what can you conclude? It may be that the Class B teacher is not very good at helping the lower-ability students, and therefore there is a wider distribution of scores at the lower end than in Class A.

> **REMEMBER:**
> All scientific calculators will work the standard deviation out for you, and you are permitted to take a calculator into the exam. To do this, you enter the individual values and use the SD function. There is no reason to ever do the calculation the long way. However, in some exam questions you may be required to show your working out – but you can check your answer using the calculator.

PRACTICE QUESTIONS

1. Work out the mean and range of the following data sets (give your answers to 2 d.p.):
   a. 5, 6, 8, 10, 12, 20, 25
   b. 5, 8, 9, 11, 12, 14, 16, 27
2. State which of the two data sets in question 1 would have the larger standard deviation.
3. A data set of 9 numbers produces a value for $\Sigma(x - \bar{x})^2$ of 62. Using the formula on the previous page (with $n - 1$), calculate the standard deviation.
4. A psychologist conducts a study of the effectiveness of two treatments for depression. Twenty patients receive therapy A and 20 patients receive therapy B. They are tested at the beginning of the study and again 6 months later. The table on the right shows the mean improvement and the standard deviation for each group. Identify two conclusions that you can draw from this data.

	Mean improvement	Standard deviation
Therapy A	7.5	5.5
Therapy B	7.1	2.5

HALFWAY SUMMARY QUESTIONS

1. In the Piliavin *et al*. (1969) study (described in more detail on page 23), researchers wanted to see whether people on a subway train would be more likely to help a confederate who collapsed if the confederate had a walking stick (appeared disabled) or carried an alcoholic drink (appeared drunk). The researchers found that in 62 of the 65 'disabled' trials people helped the confederate and in 19 out of the 38 'drunk' trials people helped the confederate.

A further type of data was collected by researchers who observed the behaviour of other passengers on the train and recorded their comments, such as 'You feel so bad when you don't know what to do'.

- **a.** Identify the sampling method used in this study to select participants. Explain the reason for the answer you choose.
- **b.** Explain why the data collected in this study would be classed as primary data.
- **c.** Give an example of quantitative data collected in this study.
- **d.** Give an example of qualitative data collected in this study.
- **e.** Explain the value of collecting both quantitative and qualitative data in this study.
- **f.** Identify the level of measurement of the data collected. Justify your answer.

2. The study by Milgram (1963), described in more detail on page 23, recruited 40 male participants by placing an advertisement in a local newspaper and sending out a mailshot. This study looked at obedience to destructive orders.

- **a.** Identify the sampling method used in this study to select participants. Explain your answer.
- **b.** Give one disadvantage of using this method of sampling in this study.
- **c.** Give one advantage of this method of sampling in this study.
- **d.** Haslam *et al*. (2014) re-analysed the way participants responded to the orders (prods) in Milgram's study. They found that all participants who were told 'You have no other choice, you must go on' stopped, i.e., they did not obey when the order was direct. Explain why the data used in Haslam *et al*.'s study would be classed as secondary data.
- **e.** Identify the level of measurement of Haslam *et al*.'s data. Justify your answer.

3. Rosenhan (1973) conducted a study on the validity of psychiatric diagnosis. He asked 8 of his acquaintances if they would present themselves as pseudopatients at various mental hospitals. They were all admitted and observed the behaviour of nurses, doctors, and the real patients by writing down any behaviours they regarded as significant.

- **a.** Identify the sampling method used in this study to select participants. Explain your answer.
- **b.** Identify the sampling method used in this study to select observations. Explain your answer.
- **c.** Identify the level of measurement of the data collected and explain your answer.
- **d.** The pseudopatients remained in the hospitals for 7 to 52 days. State the range for this data.
- **e.** Explain why you might add a correction of 1 to the range.
- **f.** The mean time spent in the hospitals was 19 days. Explain why the mean might not be the best measure of central tendency to use in this study.

4. Simons and Chabris (1999) demonstrated visual inattention in a study where participants watched two teams playing basketball. In the middle of the game a man in a gorilla suit walked between the players and stopped momentarily to beat his chest. There was an easy condition where participants just counted passes of the basketball, and a hard condition where participants had to count bounce passes and aerial passes.

   a. The researchers found that the standard deviation of the total pass counts for the easy condition was 2.71 and 6.77 for the hard condition. Explain why this supports the view that the hard condition was more difficult.

   b. The researchers decided to discard the data from participants whose scores were more than 3 standard deviations from the mean. Explain why they did this.

   c. The data in this study was largely quantitative but the key finding was qualitative – they asked participants to provide details of anything they saw in the video. Explain why this is qualitative data.

5. Baron-Cohen *et al*. (1997) investigated theory of mind as an explanation for autism by comparing three groups of participants.

   a. One group were people who responded to an ad in the *National Autistic Society* magazine. Identify the sampling method and explain why this is a good way to obtain a sample of people with autism.

   b. A second group were a random sample drawn from a database of people at the University of Cambridge. It was presumed that they had no disorders and therefore were a control group. Explain how this random sample might be obtained.

   c. The third group were participants with Tourette syndrome who acted as another control group. Explain what method of sampling was likely to be used and why.

   d. The researchers tested all participants using a range of tests, such as the Eyes Task and a gender recognition test. The results were expressed as numerical scores. Identify whether the type of data collected would be quantitative or qualitative, and primary or secondary. Explain your answer.

   e. The Eyes Task requires participants to look at pictures of eyes and for each picture identify the emotion shown by selecting one of two adjectives provided. Explain why the scores for this task might be classified as ordinal level data.

2.8 Frequency tables

A **frequency table** displays the frequency of a set of events, i.e., it shows how often each event occurred. It is a descriptive statistic because it allows you to describe and summarise a data set.

To construct a frequency table from a data set, you can use tally marks to work out the frequency for each event. Frequency tables can be used to represent any level of measurement (nominal, ordinal, interval, or ratio data).

 WORKED EXAMPLE

Here are a set of scores from a group of participants who took a memory test (with a maximum of 20 marks):

4, 11, 17, 12, 13, 10, 9, 12, 15, 16, 7, 11, 10, 9, 15, 12, 9, 7, 8, 3, 5, 12, 13, 19, 12, 13, 15, 14, 9, 8, 10, 10, 17, 12, 4, 9, 8, 14, 5, 8, 13, 14, 9, 4, 9, 11, 13, 3, 9, 11

You can calculate the mean and standard deviation of the scores using a calculator, but if you want to find the median, mode, range, or the distribution, then drawing a frequency table will help.

You count the frequency of each score by recording 1 tally mark each time the score occurs and striking through each time the count reaches 5. The **cumulative frequency** is the running total of all the tallies.

From the table you can see:

Mode = 9

Range = 19 − 3 = 16 (note that you have now made the +1 correction because there are no intermediate scores).

Median falls between the 25th and 26th place (between the scores of 10 and 11) = 10.5

Using a calculator:

Mean = 10.46

Standard deviation = 3.8 (1 d.p.)

Score on Maths test	Tally	Frequency	Cumulative frequency
3	||	2	2
4	|||	3	5
5	||	2	7
6			
7	||	2	9
8	||||	4	13
9	⊣⊦|||	8	21
10	||||	4	25
11	||||	4	29
12	⊣⊦|	6	35
13	⊣⊦	5	40
14	|||	3	43
15	|||	3	46
16	|	1	47
17	||	2	49
18			
19	|	1	50
Total frequency = 50			

 NOTE: Practice questions for this section can be found after section 2.9.

This column is needed to work out the median.

! REMEMBER: It is sometimes easier to see a pattern by grouping the data together, i.e., grouping together the scores 3 and 4, 5 and 6, etc. in this example.

2.9 Ranking data

In section 3 you will look at statistical tests for analysing the significance of research results. One of the techniques required is **ranking** data – placing data in order from smallest to largest.

If you have a large amount of data then it helps to use a **frequency table** to calculate the ranks.

WORKED EXAMPLE

When values are not repeated it is easy to rank them.

For example, the memory test scores for 1 class might be: 3, 5, 8, 9, 12, 14

The person with a score of 14 is 1st (rank position of 1) whereas the person with a score of 3 is 6th (rank position of 6).

However, if the scores were 3, 5, 8, 9, 12, 12, then 2 people would rank 1st. They would have to share rank 1 and 2 (tied rank) because the value 9 is still 3rd (rank 3). So you would then need work out the mean value of rank 1 and 2 = 1.5

The table shows an example of how tied ranks can be worked out with a large number of tied ranks. (This time you are starting the ranks from the lowest value.)

Score	Tally	Ranks	Final rank
2	|	1	1
3	||	2 and 3	2.5
4	||	4 and 5	4.5
5	ⅢⅡ	6, 7, 8, 9, and 10	8
6	||||	11, 12, 13, and 14	12.5
7	|||	15, 16, and 17	16
8	||	18 and 19	18.5
9	|||	20, 21, and 22	21
10	||	23 and 24	23.5

PRACTICE QUESTIONS

These questions are for you to practise what you have learned in sections 2.8 and 2.9

1. Here are the results of a survey that asked 13 people to name their favourite psychologist: Milgram, Zimbardo, Milgram, Bandura, Milgram, Freud, Milgram, Zimbardo, Loftus, Zimbardo, Asch, Freud, Loftus.

   a. Draw a frequency table to represent these results.

   b. Identify the mode for this data set.

   c. State the level of measurement.

   d. You initially had qualitative data from the survey and ended with quantitative data. Explain this.

2. A group of 33 students took a memory test. These are their scores out of a maximum of 20: 12, 15, 14, 3, 7, 9, 15, 13, 9, 5, 7, 9, 7, 16, 19, 5, 7, 10, 7, 3, 14, 12, 11, 9, 7, 10, 11, 11, 14, 9, 15, 14, 6

   a. Draw a frequency table to represent these results.

   b. Rank the items in this data set.

   c. Give the range, mode, and median for this data set.

   d. Redraw the frequency table but this time group the values 1–4, 5–8, 9–12, 13–16, and 17–20

2.10 Frequency diagrams: bar charts and histograms

Both **bar charts** and **histograms** are methods to illustrate a **frequency distribution**. They both use bars to illustrate the number of cases in any category.

There are several key differences between a bar chart and a histogram:

* Bar charts can be used for nominal data but histograms can't.

* Bar charts are not just used for frequency data. The y-axis can display percentage values, mean scores, and so on.

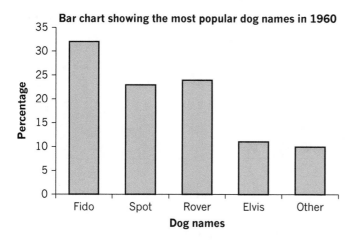

Bar chart showing the most popular dog names in 1960

> **NOTE:**
> The items on the x-axis in the bar chart on the left can be arranged in any order.

* Histograms display **continuous data**, meaning the values on the x-axis have a fixed order. For example, if you wish to display data on a bar chart about favourite names for dogs you can arrange the names in any order – you might choose to use alphabetical order but do not have to. However, if you draw a frequency diagram (histogram) of scores on a test you must use the numerical sequence (continuous data).

* In a histogram the area of the bars must be proportional to the frequencies represented. This means that the vertical and horizontal axes must start at zero, though a gap in the axis with disrupted lines (as below) can indicate missing numbers to avoid having some very tall columns or large empty spaces.

* Finally, in a bar chart there are gaps between the bars because the data are not continuous. There are no gaps in a histogram.

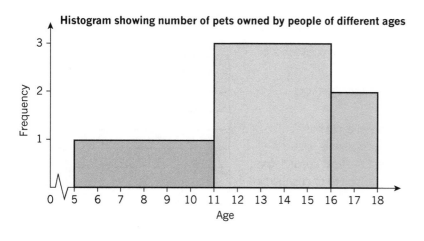

Histogram showing number of pets owned by people of different ages

> **NOTE:**
> In a histogram a break is shown if the numbering is not continuous and the area of the bars represents the frequency – in the case here, the yellow bar is 5 units × 3, which represents 15 people. In a bar chart, the length of the bar alone represents frequency.

There are two important points to note about drawing bar charts or histograms:

1. You must always label both the x and y axes clearly and have a title for the diagram.

2. You can draw the graph sideways and have horizontal bars.

WORKED EXAMPLES

1. Constructing a frequency diagram

To draw a bar chart or histogram, first construct a frequency table as you did in section 2.8. You can then turn this into a bar chart or histogram:

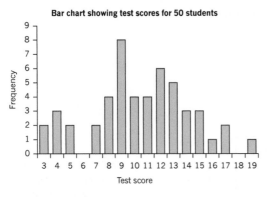

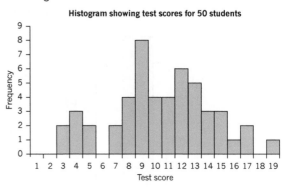

> **REMEMBER:**
> The bar chart and histogram are very similar, as is often the case – but when data are **continuous** you can't use a bar chart, and when data are **nominal** you can't use a histogram.

2. Interpreting a bar chart or histogram

An exam question may ask for conclusions to be drawn from a bar chart or histogram. This is not the same as stating the fact. For example, in the diagrams above you might comment that a mark of 9 was the most common – this is not a conclusion, it is a statement of fact.

For a conclusion, you need to indicate what this suggests about the participants' behaviour. For example, the highest values are 9, 12, and 13, which suggests that students didn't know this topic very well as very few got more than 60% correct.

PRACTICE QUESTIONS

1. Draw a bar chart for the data from practice question 2 in section 2.9

2. Identify three differences between a bar chart and a histogram.

3. A student was asked to draw a bar chart of her questionnaire results. She asked participants to name their favourite colours and coloured the bars accordingly. Identify four things that should be added to or changed in the chart on the right for it to be correct.

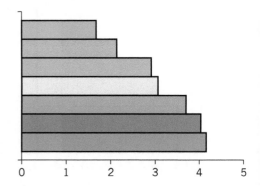

4. The diagram on the right gives the results of the study by Loftus and Palmer (1974) where participants were shown pictures of a car accident and asked 'About how fast were the cars going when they XXX each other?', where XXX is one of the verbs given.

a. State one conclusion you could you draw from the bar chart on the right.

b. Explain why a bar chart is appropriate in this case rather than a histogram.

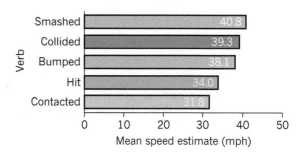

Bar chart showing the results of the Loftus and Palmer (1974) study

5. The testing effect was investigated by Roediger and Karpicke (2006). One group of participants studied a passage of text and then reread it. Another group studied the passage then tested themselves. Both groups took a recall test after 5 minutes, 2 days, and 1 week.

a. On the right is a graph showing the results of the study. State two conclusions that you could draw from this bar chart.

b. Explain why this graph is a bar chart.

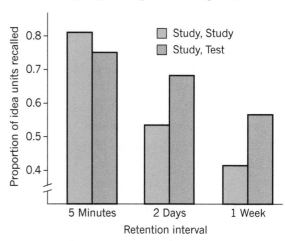

Bar chart showing results of Roediger and Karpicke (2006) study looking at the testing effect

2.11 Normal distribution

A histogram is a frequency distribution – it provides an illustration and a summary of how a data set is spread out.

There is one histogram/frequency distribution – the **normal distribution** – that is particularly important because it represents the way any data set should be distributed if you took a large enough random sample and measured physical qualities (e.g., height or the life of a lightbulb). In a normal distribution you would expect most data items to cluster around the mean, with fewer items at the two extremes.

Key characteristics

The characteristics of a normal distribution are specifically defined:

*   It forms a bell-shaped curve which is symmetrical about the midpoint.

*   The mean, median, and mode are at this midpoint.

*   The distribution of the data is specified:

    o   99.73% of the scores are three standard deviations above and below the mean (49.87% above and 49.87% below).

    o   95.44% of the scores are two standard deviations above and below the mean (47.72% above and 47.72% below).

    o   68.26% of the scores are one standard deviation above and below the mean (34.13% above and 34.13% below).

*   The tail ends never meet the horizontal axis.

*   The normal distribution has one mode – you can also have a bi-modal distribution (as shown on right) or other irregular distributions.

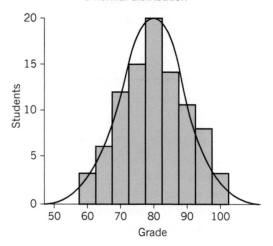

Frequency distribution approximating to a normal distribution

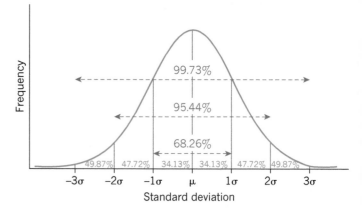

The theoretical normal distribution

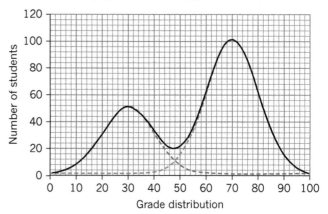

A bimodal distribution which suggests two separate populations are being tested

WORKED EXAMPLE

A psychologist recorded how long each of 100 participants took to identify a particular stimulus.

If you draw this distribution, you can see it could be described as a normal distribution. The mean, median, and mode are approximately the same, and the graph either side of the central point is roughly symmetrical.

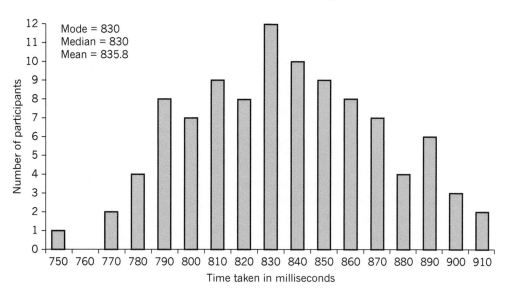

Time taken for participants to identify a stimulus

Mode = 830
Median = 830
Mean = 835.8

y-axis: Number of participants
x-axis: Time taken in milliseconds

PRACTICE QUESTIONS

1. Describe three characteristics of a normal distribution.

2. IQ test scores are normally distributed, with a mean of 100 and a standard deviation of 15

   a. State the expected percentage of the population that should have an IQ in the range of 85 to 115.

   b. State the expected percentage of the population that should have an IQ of less than 85.

3. A psychologist found that the mean number of unrelated digits a group of participants could correctly recall on a digit span test was 7. The data was normally distributed, and the standard deviation for this group was 2

   a. What percentage of participants would score less than 5?

   b. What percentage of participants would score more than 11?

   c. What percentage of participants would score between 7 and 11?

2.12 Skewed distributions

There are two other named distributions that psychology students are required to study – **skewed distributions** in a positive or negative direction.

- In a skewed distribution the mean, median, and mode do not share the same position i.e., they have different values – in a normal distribution the mean, median and mode all have the same value.

- In a **negative skew,** most of the scores are bunched to the right and some tail off in a negative direction (to the left). Most of the values lie at the peak which is the mode. The mode is therefore to the right of the mean and the median. The mean has the lowest value as it is most affected by the scores in the tail.

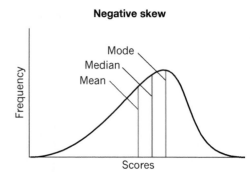

- In a **positive skew,** most of the scores are bunched towards the left and some tail off in a positive direction (to the right). The mode is therefore to the left of the mean and the median. The mean has the highest value.

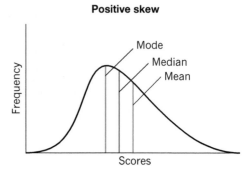

 WORKED EXAMPLE

When psychologists design a test they need to avoid floor and ceiling effects.

A floor effect occurs when a test is too difficult and therefore most people get very low scores. For example, the data might look like this for a test with 20 items: 2, 2, 3, 4, 4, 4, 4, 5, 6, 6, 6, 8, 10, 11, 11, 13, 15 If you draw this distribution you can see that it could be described as positively skewed.

> **REMEMBER:**
> Real data are unlikely to be an exact fit for any type of distribution – you are looking at the general pattern.

The mode = 4, median = 6, mean = 6.7

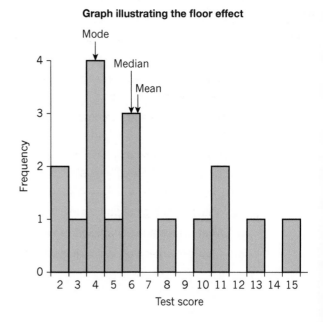

A ceiling effect occurs when a test is too easy so most people get very high scores.
For example: 2, 5, 6, 8, 10, 12, 13, 13, 14, 15, 15, 15, 15, 16, 16, 17, 20

The mode = 15, median = 14, mean = 12.5

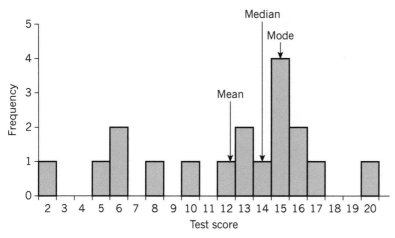

Graph illustrating the ceiling effect

PRACTICE QUESTIONS

1. Answer the following questions:

   a. Explain the difference between a normal and a skewed distribution.

   b. Sketch a distribution that is positively skewed and indicate the position of the mean, median, and mode.

   c. Describe the characteristics of a negatively skewed distribution.

2. One method that is used to diagnose depression is to give a person a checklist of symptoms, such as 'My future feels hopeless' and 'The pleasure has gone out of my life.' A high score on this checklist indicates depression. Identify the likely distribution of scores if 100 people who report feeling depressed were tested. Explain your answer.

3. A teacher gave her students a psychology test to see how much they had understood of the topic they had just finished. She found that most of her students scored very highly on the test.

   a. Identify what type of distribution she would see if she drew a distribution graph of their results.

   b. Explain why her results have produced this type of distribution.

   c. Explain how she could modify her test so that it will produce a spread which is more similar to a normal distribution.

2.13 Scatter diagrams

Scatter diagrams are a type of graph that is specifically used to represent correlations.

A correlational analysis is a way of considering the relationship between two continuous co-variables (co-variable A and co-variable B). For example, you can look at the relationship between birth order and intelligence, or time spent asleep and time spent exercising in a 24-hour period.

A simple visual analysis of the relationship between co-variables can be made using descriptive rather than inferential statistics – by drawing a scatter diagram.
The pattern of dots indicates the type and strength of the relationship between co-variables.

A scatter diagram is drawn by:

* Labelling each axis with one of the co-variables.
* For each case/participant in the sample, locate the correct value for co-variable A and co-variable B and then plot a point.

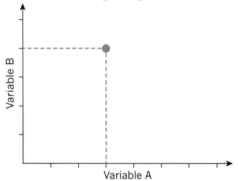

> **REMEMBER:**
> Correlations never demonstrate a cause. You cannot conclude that one of the variables *caused* a change in the other variable, you can only conclude that the two are related. This is because there may be **intervening variables** that could explain the relationship.

You will this look at how correlational analysis can be measured more precisely in section 3 on inferential statistics (see sections 3.11 and 3.12).

 WORKED EXAMPLES

Consider the relationship between stress and performance. You can measure each of these variables on a continuous scale. Stress could be measured by giving participants a questionnaire to assess their stress levels, and performance could be measured by giving participants a task to complete and scoring how well they do. Thus, you have operationalised your two variables and for each participant you will have two scores (continuous data).

You can plot the scores for 20 participants with the x-axis being stress and the y-axis performance. Your data would fit one of the six patterns shown on the next page.

A perfect positive correlation is represented as $+1.0$ (this is called the **correlation coefficient**), in which case all the dots would lie on the diagonal straight line. A strong correlation would be a value above $+.50$

Strong positive correlation

Strong positive correlation: Points are arranged from bottom left to top right illustrating that as co-variable A increases so does co-variable B. The closer the dots are to a diagonal straight line the more the two co-variables are related. In this case, the scatter diagram suggests that as stress increases, performance also increases. So, you might conclude that people with high levels of stress tend to perform well at the task.

Moderate positive correlation

Moderate positive correlation: Points are less clearly in a diagonal straight line but still show a tendency to go from bottom left to top right. The correlation coefficient for a moderate positive correlation (calculated using a statistical test) would be between about +0.30 and +0.50

No correlation

No correlation: No discernible pattern. This is sometimes referred to as zero correlation.

Strong negative correlation

Strong negative correlation: Points are arranged from top left to bottom right, suggesting that as co-variable A increases, co-variable B decreases. Again, the closer the points are to a diagonal straight line, the more closely the co-variables are related. This scatter diagram suggests that as stress increases, performance decreases. So, you might conclude that people with high levels of stress tend to perform less well on this task. A perfect negative correlation is represented as −1.0

Moderate negative correlation

Moderate negative correlation: Points are less clearly in a diagonal straight line but still show a tendency to go from top left to bottom right. The values for a moderate negative correlation would be between about −0.30 and −0.50

 NOTE:
−0.50 is as strong a correlation as +0.50

The sign just shows the direction of the relationship between co-variables, not the strength.

Curvilinear relationship

Curvilinear relationship: In some cases there is not a consistent increase or decrease in the points, and the relationship may be curvilinear rather than linear (straight line). This scatter diagram would suggest that both low and high levels of stress are related to good performance whereas moderate levels of stress are related to poor performance.

PRACTICE QUESTIONS

1. Sketch a scatter diagram for the following correlations:
   a. Moderate negative correlation.
   b. Correlation of +1.0
   c. Zero correlation.
   d. Strong positive correlation.

2. Describe the kind of scatter diagram you would draw for the following:
   a. A correlation of −.30
   b. A correlation of −.80
   c. A correlation of +.50
   d. A correlation of +.10

3. Giuseppe Gelato always liked statistics at school and now that he has his own ice cream business he keeps various records. The table below shows his data.

	Jan	Feb	Mar	Apr	May	Jun	Jul	Aug	Sep	Oct	Nov	Dec
Ice cream sales	10	8	7	21	32	56	130	141	84	32	11	6
Aggressive crimes	21	32	29	35	44	55	111	129	99	36	22	25

All data rounded to 1000s

   a. Sketch a scatter diagram of Giuseppe's data.
   b. What can you conclude from the data and the scatter diagram?
   c. With reference to your conclusion in part b, what intervening variable might explain the relationship between ice cream sales and aggressive crimes?

SUMMARY QUESTIONS FOR SECTION 2

1. A researcher measured the time taken (in seconds) for a group of children to complete a puzzle. The scores collected were:

   10, 12, 25, 19, 12, 20, 16, 18, 22, 18, 17, 26, 21, 19, 15, 18, 17, 15, 16, 24, 9, 28, 18, 13, 21, 11, 20, 24, 19, 20, 16, 22

   a. Draw a frequency table for the data.

   b. Rank the data and use these to work out the median.

   c. If the data are normally distributed, state what the value of the mean and mode should be.

   d. Calculate the mean and mode for the group of data.

   e. Explain why a histogram would be a suitable frequency diagram for the data.

   f. Draw a histogram of the data.

   g. Explain why this distribution might be described as a normal distribution.

   h. Sketch a normal distribution.

   i. A second group of participants did the same task and, when the data was plotted, it showed a positively skewed distribution. Sketch what the distribution would look like.

   j. If the mean for the second group was 16, would the median be larger or smaller? Explain your answer.

   k. A third group of students produced a negatively skewed distribution. Explain the difference between data that is positively and negatively skewed with reference to these second and third samples of data.

2. Psychological research has found that partners in a married couple each have a similar rating in terms of attractiveness.

   a. Describe the kind of correlation you would expect to find based on the prediction that partners in a married couple tend to have a similar rating in terms of attractiveness.

   b. To conduct this study researchers assembled photographs of couples and showed them to a sample of participants who rated them. Suggest a suitable method of obtaining this sample of participants.

   c. The data below are attractiveness scores for 15 couples. Sketch a scatter diagram using the data.

	1	2	3	4	5	6	7	8	9	10	11	12	13	14	15
Partner 1	10	5	6	9	3	4	8	9	8	4	6	4	7	6	3
Partner 2	8	6	9	7	5	4	7	9	9	6	6	5	6	7	4

   d. Identify the type of data in this study.

   e. Estimate the correlation coefficient for the data shown in the scatter diagram.

   f. Researchers have found that this relationship does not hold true in cases where people have known each other for a long time before becoming a couple. If a psychologist tested the ratings of both partners who had been friends for a long time before becoming a couple, identify the correlation that should be found.

   g. Explain the difference between a positive and a negative correlation, using the relationship between different kinds of couple as an example.

3. Maguire *et al*. (2000) investigated the brains of taxi drivers, looking at the size of their hippocampus – an area of the brain linked to spatial memory. They found that the hippocampus was larger in taxi drivers compared to control participants. The control participants were men who were not taxi drivers, and were drawn from a group of 50 men whose records were held on a database.

   a. Identify the primary data and secondary data in this study.

   b. Maguire *et al*. found a moderate positive correlation between time spent as a taxi driver and the volume of the posterior right hippocampus. State a conclusion you could draw from this.

   c. A graph was drawn to show the relative volume of three areas of the hippocampus (anterior, middle or body, and posterior).

   Explain why this is a bar chart and not a histogram.

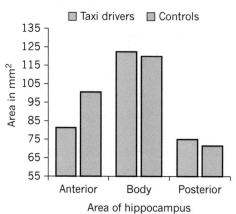

Hippocampal volume bar chart

4. Elms (2000) tested some of the participants in Milgram's obedience studies, looking at how obedient they were in the study and their scores for having an authoritarian personality (the extent to which a person believes in being submissive to people in authority).

   a. Elms found a positive correlation between the shock level delivered and a person's score for having an authoritarian personality. Explain what this means.

   b. Explain why the results do not suggest that having an authoritarian personality causes greater obedience.

3 INFERENTIAL STATISTICS

3.1 Simple probability and the null hypothesis

When you talk about probability, you are talking about how probable (or likely) it is that a certain event will happen. For example:

- The probability of getting heads when you toss a coin is 1 in 2 (0.5).
- The probability of getting a 2 when you roll a six-sided dice is 1 in 6 (0.167).
- The probability of drawing a playing card with hearts or diamonds from a normal deck of cards is 26 in 52, which can be reduced to 1 in 2 (0.5).

Probabilities are calculated by dividing the outcome(s) of interest by the number of all possible outcomes. Probability is given as a number between 0 and 1 (where 0 indicates impossibility and 1 indicates certainty).

Probability can be expressed in fractional form (e.g., 2 in 6) and reduced to the simplest form (e.g., 2 in 6 is the same as 1 in 3), or can be expressed as a decimal fraction (e.g., 0.33).

Why probability matters in Psychology

The interest in probability as psychologists is to know the meaning of the results from a research study — what do the results tell you about the population? To do this you use **inferential statistical tests** and these statistics are based on the mathematics of probability.

Consider the following: If you wanted to know whether women are better at being leaders than men you could not test all of the women and men in the world (the total population), so you just test a small group of women and a small group of men (i.e., you test a **sample** of the target population).

If you find that the sample of women are indeed better at being leaders than the sample of men, the question is whether this is true of the whole population. It is possible that the difference you found simply occurred by chance and there is no 'real' difference between women and men.

You cannot be 100% certain that the observed effect was not due to chance. However, what you can do is state how certain you are that the result is 'real'.

The null hypothesis

In order to determine whether the difference in your sample is likely to have occurred even if there is no real effect in the population, psychologists refer to the **null** hypothesis (abbreviated as H_0). This is a statement of no effect (nothing is going on). In other words, there is no difference or correlation. In your hypothetical investigation the null hypothesis would be 'There is no difference in leadership ability between men and women'.

Psychologists then use inferential statistical tests to work out, at a given probability, whether the null hypothesis can be reasonably rejected. If you reject the null hypothesis, you can accept the **alternative hypothesis** (abbreviated as H_1), such as 'There is a difference in leadership ability between men and women', or you could specify the kind of difference 'Women are better at being leaders than men'.

You will look at these inferential statistical tests in section 3.4 onwards.

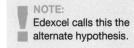
NOTE:
Edexcel calls this the alternate hypothesis.

Level of probability

In general, psychologists use a level of probability of 0.05. This level of probability expresses your degree of certainty. It means there is a probability of 0.05 (a 5% chance) of getting the observed results if the null hypothesis was true. In other words, there is a 5% chance that the results would occur even if there was no real difference/correlation between the populations from which the samples were drawn.

This probability of 0.05 is recorded as $p = 0.05$, where p means probability. In fact, it is more accurate to say a probability equal to or less than 0.05, which is written as $p \leq 0.05$

In some studies psychologists want to be more certain (such as when they are testing the effects of a new drug). Then, researchers use a more stringent probability, such as $p \leq 0.01$ or even $p \leq 0.001$

These probabilities are also expressed as significance levels. A probability of 0.01 is a 1% significance level (or 1% chance) that an event occurred due to chance. A probability of 0.001 is a 0.1% significance level (or 0.1% chance) that an event occurred due to chance.

WORKED EXAMPLE

A teacher has a class of 30 pupils. 18 of these are boys and 12 are girls. The probability of a pupil being a boy is 18 out of 30 (60%) and the probability of a pupil being a girl is 12 out of 30 (40%).

PRACTICE QUESTIONS

1. There are 30 balls in a basket, 6 are red. A child draws out a ball with their eyes blindfolded. State the probability of drawing a red ball from the basket.

2. Explain what a probability of 0.05 means.

3. For each of the following change the decimal fraction to a percentage:

    a. 0.01

    b. 0.10

    c. 0.05

    d. 0.001

    e. 0.03

4. A student claims they can correctly identify the suit of a randomly selected card from a pack of 52 cards.

    a. What is the probability of the student correctly identifying the suit of a selected card?

    b. The student predicts that the first card drawn will be a Jack, Queen, or King. What is the probability of the student being correct?

    c. The student then changes their mind and predicts that the first card will be a red seven. What is the probability of the student being correct?

3.2 Hypothesis testing

Section 3.1 introduced the two types of hypothesis: **alternative** (or alternate) and **null hypotheses**. In theory, these are both stated at the beginning of a research project. However, most research papers actually give a research *prediction*. This is a prediction of what the researcher(s) expect *will* be the outcome of their study for their sample participants.

By contrast a hypothesis is a statement of what the researcher believes to be true for the *population* (not just the sample). It should be given in the present tense and the word 'significant' should not be used because only inferential statistical tests assess significance.

The alternative hypothesis is a statement of effect (something is going on), which may concern a difference or a correlation between two samples. (More complex research looks at more than two samples at a time.)

Directional and non-directional hypotheses

Alternative hypotheses can be written as directional or non-directional:

- A **directional hypothesis** states that there is a difference or correlation, and specifies which direction the difference or correlation is.

- A **non-directional hypothesis** states that there is a difference or correlation, but doesn't specify which direction the difference or correlation is.

Directional hypotheses are usually chosen when theory or past research suggests that one group differs in a specified way to another group. If past research is contradictory, or there is no past research, then a non-directional hypothesis should be used.

WORKED EXAMPLES

Hypotheses can be written for a difference or a correlation.

Example 1

In the case of a difference, these are examples of possible hypotheses:

- Directional alternative hypotheses:

  Women are *better* at being leaders than men.

  Group 1 is *slower* at finishing a task than group 2.

- Non-directional alternative hypotheses:

  Women are *different* at being leaders than men (you have not said how the results might differ; just that you think there is a difference.)

  The mean score for group 1 is different to the mean score for group 2.

Example 2

In the case of a correlation, these are examples of possible hypotheses:

- Directional alternative hypotheses:

  There is a positive correlation between height and IQ test score.

  There is a negative correlation between age and memory test score.

- Non-directional alternative hypotheses:

  There is a correlation between height and IQ test score.

  There is a correlation between age and memory test score.

Example 3

In null hypotheses no direction is stated (it is always non-directional). For example:

There is no difference between men and women in terms of leadership ability.

Women are not different at being leaders to men.

The mean score for group 1 is not different to the mean score for group 2.

There is no correlation between height and IQ test score.

There is no correlation between age and memory test score.

PRACTICE QUESTIONS

1. A psychologist believes that there would be a difference in the number of times male and female drivers fail to stop at a zebra crossing. Is this a directional or non-directional hypothesis?

2. Explain why the hypothesis that drinking two units of alcohol decreases participants' reaction times is directional, rather than non-directional.

3. A researcher hypothesised that people solve difficult problems more slowly when they are observed by others compared to when they are alone.

   a. Is this a directional or non-directional hypothesis? Explain your answer.

   b. Write a corresponding null hypothesis for this study.

4. A researcher writes an alternative hypothesis for a study: 'People are able to solve more anagrams in silence than when music is playing'. Another researcher disagrees and suggests this is a better hypothesis: 'People are able to solve fewer anagrams in silence than when music is playing'.

   a. Explain why the second researcher's hypothesis is not a null hypothesis.

   b. Write a suitable null hypothesis for this study.

3.3 Error types

Whenever an inferential statistical test is used to see if a difference or correlation is real there is a chance of making a mistake.

One way of making a mistake is to conclude that a difference (or correlation) is significant when, in the actual population, there is no difference/correlation. This means that the null hypothesis is rejected when it should not have been and is called a **Type 1 error**.

The other way of making a mistake is to conclude that a difference (or correlation) is not significant when, in the actual population, there is a difference/correlation. This means that the null hypothesis will be accepted when it should not have been and is called a **Type 2 error**.

Type 1 and Type 2 errors and significance levels

The main way a Type 1 error is made is if researchers use a **level of significance** that is too lenient. For example, using the 10% significance level ($p \leq 0.1$) instead of the conventional 5% significance level ($p \leq 0.05$).

The main way a Type 2 error is made is if researchers use a level of significance that is too stringent. For example, using the 1% significance level ($p \leq 0.01$) instead of the conventional 5% significance level ($p \leq 0.05$).

The 5% significance level is usually used by psychologists as it is seen as being an acceptable balance between the risk of making a Type 1 and a Type 2 error (i.e., avoiding being too lenient or too stringent).

> **NOTE:**
> The 5% compromise is accepted as a reasonable answer in A level exams. However, in actual research situations, risk is also an important factor, and therefore many researchers select 1% ($p < 0.01$) to err on the side of caution.

WORKED EXAMPLES

Example 1

You will consider a real world example of a jury trial. The null hypothesis is that the defendant is not guilty (nothing happening, he is innocent). The alternative hypothesis is that the defendant is guilty. Possible significance levels you could use:

- 0.01 probability (1% significance level). This is very stringent. Being stringent means you may *accept* a null hypothesis that is in fact false (a false negative). You accept the null hypothesis that the defendant is not guilty and therefore let him go free. The reality (which you will never know) is that he is guilty and you have let a guilty man go free. This is a Type 2 error.

- 0.10 probability (10% significance level). This is very lenient. Being lenient means you may *reject* a null hypothesis that is in fact true (a false positive). You reject the null hypothesis that the defendant is not guilty and therefore put him in prison. The reality (which you will never know) is that he is not guilty and you have put an innocent man in prison. This is a Type 1 error.

- 0.05 probability (5% significance level). This is a compromise between being too stringent (reluctant to reject the null hypothesis) and too lenient (too willing to accept the alternative hypothesis).

> **NOTE:**
> False negative means accepting something as not true (i.e. negative) that in fact is true.
>
> False positive means accepting something as true (i.e. a positive identification) that in fact is not true.

> **NOTE:**
> One way of remembering what a Type 2 error means is the acronym for 'Ta ta for now' (a quaint British farewell):
>
> TTFN = Type Two False Negative – falsely accepting the null hypothesis.

Example 2

You can consider the same arguments in the context of the leadership example from earlier (see section 3.1).

- 0.01 probability (1% significance level). You risk wrongly *accepting* the null hypothesis when it is false. This means you overlook a real difference between men and women, a Type 2 error.

- 0.10 (10% significance level). You risk wrongly *rejecting* the null hypothesis. This means you accept a difference between men and women that doesn't exist, a Type 1 error.

Example 3

A psychologist wishes to test the hypothesis that older people and younger people differ in the accuracy of their eyewitness testimony. The null hypothesis is that there is no difference between older and younger people. The alternative hypothesis is that there is a difference between them.

- If you use a 1% significance level, the risk of a Type 2 error is *more* likely than a Type 1 error. You risk accepting a null hypothesis that is false and missing a true effect. In this case, you accept that there is no difference between older and younger people in terms of their eyewitness testimony accuracy.

- If you use a 10% significance level, the risk of a Type 2 error is *less* likely than a Type 1 error. You risk rejecting a null hypothesis that is true and accepting a false effect. In this case, you accept that there is a difference between older and younger people in terms of their eyewitness testimony accuracy.

PRACTICE QUESTIONS

1. Explain the difference between using the 0.05 and 0.01 levels of significance.

2. Psychologists may use the 0.05 significance level when deciding whether to accept or reject a null hypothesis. Explain why.

3. A psychologist uses a 5% level of significance in a research study.

    a. Give the likelihood of making a Type 1 error in this study.

    b. What would the likelihood have been of making a Type 1 error if the psychologist had used the 1% level of significance?

    c. Is a Type 2 error more or less likely when the 1% significance level is used, rather than the 5% significance level? Explain your answer.

3.4 Using inferential statistical tests

The terms statistical test, inferential test, inferential statistics, or even inferential statistical test are used interchangeably. You will use the phrase 'inferential test'.

There are two key points to remember:

- There are two groups of statistics – descriptive and inferential. You covered descriptive statistics in section 2.

- Inferential tests enable you to draw inferences (conclusions) about populations from the samples of data tested. The methods are based on the probability of the sample data occurring if the same difference/correlation was true in the population.

Descriptive statistics can be used when considering the meaning of a data set because they provide a summary of the data. Descriptive statistics help you detect general patterns and trends. However, strictly speaking you cannot actually draw conclusions from descriptive statistics because you cannot go beyond the particular sample to draw inferences about people in general (the 'population').

Inferential tests allow you to make inferences beyond the particular sample so you can say, for example, 'This data suggests that all women are better leaders than men'. However, you can never state this with absolute certainty — you must state the probability or certainty that this inference is true. This is also referred to as the level of probability or level of significance (see section 3.1).

WORKED EXAMPLE

In sections 3.5–3.12 you will look at the inferential tests required in the A level psychology specifications. There are certain steps that are followed for all of these tests:

Step 1: Decide which test to use.

This depends on three factors:

- Is a test of difference or association/correlation required?
- If a test of difference, is the experimental design independent groups or repeated measures/matched pairs?
- What is the type of data: nominal, ordinal, or interval? (see section 2.6)

	Test of difference		Test of association or correlation
	Independent groups	**Repeated measures**	
Nominal	Chi-squared (χ^2)	Sign test (S)	Chi-squared (χ^2)
Ordinal or better	Mann–Whitney (U)	Wilcoxon (T)	Spearman's (rho)
Interval data	Unrelated t-test (t)	Related t-test (t)	Pearson's (r)

 NOTE:
Spearman's and Pearson's tell you the strength of the correlation between two variables (the correlation coefficient), and whether the correlation coefficient is significant or not. Chi-squared does not produce a correlation coefficient. Rather, it tells you if there is an association between two independent variables. For example, a chi-squared test could tell you if there was an association between gender and driving ability. This is not the same as measuring the correlation between those variables.

Step 2: State the hypotheses (alternative and null).

Step 3: Record data and preliminary calculations.

Step 4: Find the calculated value. The value that is worked out is called the **calculated value** (or observed value). Each inferential test has a formula for doing this that produces a single number called the test statistic (see the table in step 1 for inferential test abbreviations). The methods of test calculation are described in the rest of section 3.

Step 5: Is the result in the right direction? If the alternative hypothesis is directional (one-tailed) and the results are *not* in the stated direction then *the null hypothesis must be accepted* even if the calculated value is significant.

Step 6: Find the critical value. This is the value needed in order for the null hypothesis to be rejected. For each test there is a specific critical values table that needs to be used (e.g. see Wilcoxon table of critical values on right).

To find the **critical value** you need to know the following information:

- Is a one-tailed or two-tailed test required? If the hypothesis is directional, a **one-tailed test** is used. For a non-directional hypothesis, a **two-tailed test** is used.

- The significance level selected, usually $p \leq 0.05$ (5% level).

- The degrees of freedom (*df*) or *N* (number of participants). This varies according to the test.

Now you use the critical values table and locate the correct orange row (one- or two-tailed) and the correct column (e.g. 0.05 for 5% level). Then locate the row that starts with your *N/df* value. The intersection of column and row gives you the appropriate critical value.

Step 7: Accept or reject the null hypothesis and draw a conclusion. At the bottom of each critical values table you will find a statement about whether the calculated (or observed) value needs to be greater than or less than the critical value for the result to be considered **significant**.

Table of critical values for the Wilcoxon test.

One-tailed test	0.05	0.025	0.01	0.005
Two-tailed test	0.1	0.05	0.02	0.01
df				
5	0			
6	2	0		
7	3	2	0	
8	5	3	1	0
9	8	5	3	1
10	10	8	5	3
11	13	10	7	5
12	17	13	9	7
13	21	17	12	9
14	25	21	15	12
15	30	25	19	15
16	35	29	23	19
17	41	34	27	23
18	47	40	32	27
19	53	46	37	32
20	60	52	43	37
25	100	89	76	68
30	151	137	120	109

Calculated value of *T* must be equal to or less than the critical value in this table for significance to be shown.

PRACTICE QUESTIONS

1. Explain the difference between using a descriptive statistic and an inferential statistic.

2. What is the name of the test statistic for Spearman's test?

3. Identify the 6 pieces of information that are needed to find whether a calculated (observed) value is significant.

4. Each study below used the Wilcoxon test. State whether the null hypothesis should be rejected.

   a. Directional hypothesis, significance level 1%, 8 participants ($N = 8$), calculated value = 4

   b. Non-directional hypothesis, significance level 10%, 20 participants ($N = 20$), calculated value = 48

3.5 Sign test

This inferential test is used when looking at differences between **paired** or **related** data. Two related pieces of data could come from a repeated measures design (i.e. the same person is tested twice). Or could come from a matched pairs design because the participants are paired and therefore count, for the purpose of this inferential test, as one person tested twice.

The test can be used with nominal data or ordinal/interval data. If nominal data are used then a $+$ would represent identical pairs of data and $-$ would represent non-identical pairs.

NOTE:
Eight inferential tests are covered in sections 3.5–3.12. Your specification may not require you to be familiar with all of these. Check your specification.

WORKED EXAMPLE

To calculate the sign test, let's imagine a study where you wanted to know whether people were happier after they had a holiday than before. To do this you ask people after their holiday to say if they were happier than before they went on holiday.

Step 1: Reasons for choosing the sign test for this study.

The hypothesis states a *difference* between two sets of data.

The two sets of data are pairs of scores from one person = *related data*.

The data are *nominal* (categorical) because you just record whether a person was happier or not.

Step 2: State the hypotheses.

Alternative hypothesis: People are happier after going on holiday than they were beforehand (a directional hypothesis). Null hypothesis: There is no difference in the happiness ratings of people from before or after their holiday.

Step 3: Record data and signs.

Record a plus ($+$) if happier after and a minus ($-$) if not happier after. This has been done in the table on the right.

Step 4: Find the calculated value of S (the symbol for the sign test).

It is calculated by adding up the pluses, adding up the minuses, and selecting the smaller value. In this case there are 10 pluses, 3 minuses, and 1 zero. Therefore, the less frequent sign is minus, so $S = 3$. This is the calculated value.

Step 5: Is the result in the right direction?

As the hypothesis is directional you have to check that the result is in the expected direction. In this scenario you expect people to be happier afterwards and should therefore have more pluses than minuses — this was the case.

Data table

Participant	Happier or not	Sign
1	Happier	+
2	Happier	+
3	Happier	+
4	Not	−
5	Happier	+
6	Not	−
7	Happier	+
8	Happier	+
9	Happier	+
10	Not	−
11	No difference	0
12	Happier	+
13	Happier	+
14	Happier	+

Step 6: Find the critical value of *S*.

- One-tailed or two-tailed test? The alternative hypothesis is directional so a one-tailed test is used.
- Significance level selected = $p \leq 0.05$ (5% level).
- Degrees of freedom (*df*) or N = 13 (any zero values are ignored).

Now you use the table of critical values and locate the column headed 0.05 for a one-tailed test and the row that begins with your N (*df*) value of 13. The critical value of $S = 3$.

Step 7: Accept or reject the null hypothesis and draw a conclusion.

For the sign test the calculated (observed) value must be equal to or less than the critical value for your result to be significant.

The calculated value (3) is equal to the critical value (3) so your result is significant at $p \leq 0.05$ (5% significance level).

This means you can reject the null hypothesis and conclude that people are happier after going on holiday than they were beforehand at $p \leq 0.05$

Table of critical values for the sign test

One-tailed test	0.05	0.025	0.01	0.005
Two-tailed test	0.1	0.05	0.02	0.01
df				
5	0			
6	0	0		
7	0	0	0	
8	1	0	0	
9	1	1	0	
10	1	1	0	0
11	2	1	1	0
12	2	2	1	0
13	3	2	1	0
14	3	2	2	1
15	3	3	2	1
16	4	3	2	1
17	4	4	3	1
18	5	4	3	2
19	5	4	4	2
20	5	5	4	2

Calculated value of *S* must be equal to or less than the critical value in this table for significance to be shown.

PRACTICE QUESTIONS

1. Convert 1% to a decimal fraction.

2. In a study involving 18 participants (with one zero score) and a one-tailed hypothesis the calculated *S* value was 5. State whether the result is significant at the 5% level.

3. The alternative hypothesis for a research study is 'There is a difference in the colours selected by twin pairs.'

   a. State the null hypothesis for this study.

   b. Fill in the correct signs in the table below, and then work out the *S* value for the pairs of scores.

Pair	1	2	3	4	5	6	7	8	9
Colour	Red	Green	Red	Blue	Blue	Orange	Purple	Orange	Red
	Blue	Green	Red	Blue	Green	Orange	Blue	Green	Orange
Sign									

   c. State the conclusion you would draw.

3.6 Wilcoxon test

This inferential test is used when looking at **paired** or **related** data. The two related pieces of data could come from a repeated measures design (i.e. the same person is tested twice), or from a matched pairs design because the participants are paired (they count, for the purpose of statistics, as one person tested twice). This test can be used with ordinal/interval data.

Non-parametric tests

The Wilcoxon test is a **non-parametric** test. Some inferential tests are described as **parametric** because the test is based on underlying assumptions (called parameters) about the population (see section 3.8).

The value of parametric tests is they have greater power than non-parametric tests. This means that if a non-parametric test does not find a significant effect, a parametric test may do so because it has increased power (like a magnifying glass). In non-parametric tests, such as the Wilcoxon test, ranking the data means that the finer details are lost. Non–parametric tests should be used with ordinal or interval level data.

WORKED EXAMPLE

Research has shown that people prefer familiar to unfamiliar things. One way to investigate this is to show participants 20 photographs of people's faces. One face appears frequently (5 times in the 20 photographs). Another face appears infrequently (1 time in the 20 photos). Other faces appear 2 or 3 times. Participants are asked to rate the attractiveness of the very frequent and very infrequent face.

Step 1: Reasons for choosing the Wilcoxon test for this study.

The hypothesis states a *difference* between two sets of data.

The two sets of data are pairs of scores from one person = *related*.

The data are *ordinal* because there are not equal intervals between ratings of the photographs.

 REMEMBER:
If you are asked to justify the test you have chosen, this justification must always be tailored as far as possible to the particular features of the study, such as explaining why the data are ordinal.

Step 2: State the hypotheses.

Alternative hypothesis: People rate the frequently seen face as more likeable than the infrequently seen face (directional hypothesis).

Null hypothesis: There is no difference in the likeability score for more and less familiar faces.

Step 3: Record the data and calculate the differences and ranks.

You need to make a table in which to calculate the differences and ranks (see next page).

- Columns 1 and 2 are the raw data – the likeability scores for each condition. There are two scores from each participant.

- Column 3 shows the difference between each score for each participant. If the difference is zero omit this from the ranking and reduce N accordingly.

- Column 4 shows the ranks for the differences – ignore the signs when ranking. Rank all data items jointly (see section 2.9). You can rank the data using a frequency table – this makes it easier when there are many tied ranks.

Calculation table

Participant	Column 1 Most frequent face	Column 2 Least frequent face	Column 3 Calculate difference	Column 4 Rank data
1	5	2	3	7
2	7	3	4	9.5
3	3	3	0	
4	8	4	4	9.5
5	2	3	−1	2.5
6	4	5	−1	2.5
7	5	2	3	7
8	3	4	−1	2.5
9	6	3	3	7
10	3	5	−2	5
11	7	2	5	11
12	3	4	−1	2.5

Step 4: Find the calculated value of T.

Add up the ranks for positive differences:
$7 + 9.5 + 9.5 + 7 + 7 + 11 = 51$
Add up the ranks for negative differences:
$2.5 + 2.5 + 2.5 + 5 + 2.5 = 15$
T is the smaller value $= 15$

Step 5: Is the result in the right direction?

The expected direction was that the more frequent face would be more likeable, so the result is in the right direction (the larger value is for the positive differences).

Step 6: Find the critical value of T.

- One-tailed or two-tailed test? The hypothesis is directional so a one-tailed test is used.
- Significance level $= p \leq 0.05$ (5% level).
- Degrees of freedom (df) or N value (total number of scores ignoring zero values). In your case $N = 11$ (1 score omitted).

Now you use the table of critical values (see section 3.4) and locate the column headed 0.05 for a one-tailed test and the row that begins with your N value of 11 – this gives a critical value of $T = 13$).

Step 7: Accept or reject the null hypothesis and draw a conclusion.

For the Wilcoxon test the calculated (observed) value must be equal to or less than the critical value for your result to be significant. The calculated value (15) is not equal to or less than the critical value (13) so your result is not significant at $p \leq 0.05$ (5% significance level).

This means you must accept the null hypothesis, and conclude that there is no difference in the likeability score for more and less familiar faces ($p \leq 0.05$).

NOTE:
Practice questions for this section can be found after section 3.7.

3.7 Mann–Whitney test

This inferential test is used when looking at **independent** or **unrelated** data. The two unrelated pieces of data are likely to come from an experiment with an independent groups design. This means the data are not paired and the number of participants in each group (N) may be different.

This test should be used with ordinal/interval data. It is non-parametric.

> **REMEMBER:**
> Here's a way to remember when to use Mann–Whitney and when to use Wilcoxon.
>
> Imagine two couples: Mr Mann and Ms Whitney, and Mr and Mrs Wilcoxon. The Wilcoxon's are related (through marriage) – the Wilcoxon test is for related data.

WORKED EXAMPLE

To calculate the Mann–Whitney test, let's look at an example. One hypothesis offered as to why people fall in love is that physiological arousal gets mislabelled as love. One way to test this is to arrange for male participants to rate a photograph of a woman for attractiveness. One group must run on the spot beforehand for 2 minutes (simulating high arousal) and the other group run for 15 seconds (low arousal). Afterwards participants are asked to rate the photograph.

Step 1: Reasons for choosing the Mann–Whitney test for this study.

The hypothesis states a *difference* between two sets of data.
The two sets of data are *unrelated* as an independent groups design is used.
The data are *ordinal* because there are not equal intervals between ratings of attractiveness.

Step 2: State the hypotheses.

Alternative hypothesis: People who run for 2 minutes (high arousal) rate a woman as more attractive than people who run for 15 seconds (low arousal) (a directional hypothesis).
Null hypothesis: There is no difference in the attractiveness ratings for the high and low arousal conditions.

Step 3: Record the data and calculate differences and ranks.

- The data for Group A: 7, 10, 8, 6, 5, 8, 9, 7, 10, 9, and for Group B: 4, 6, 2, 5, 3, 5, 6, 4, 5, 7, 9, 3, 5, 6

- Rank all of the data items jointly (see how to rank data in section 2.9). You have ranked the data (see Table 1) using a frequency table. This makes it easier when there are many tied scores.

Table 1: Frequency table to calculate ranks

Score	Frequency	Ranks	Final rank
2	1	1	1
3	11	2 and 3	2.5
4	11	4 and 5	4.5
5	~~1111~~	6, 7, 8, 9, and 10	8
6	1111	11, 12, 13, and 14	12.5
7	111	15, 16, and 17	16
8	11	18 and 19	18.5
9	111	20, 21, and 22	21
10	11	23 and 24	23.5

- Column 2 Table 2 shows the ranks for Group A data (R_A) and Column 4 shows the ranks for Group B data (R_B).
- At the bottom of Table 2 are the totals for all Group A ranks (ΣR_A) and all Group B ranks (ΣR_B).

Step 4: Find calculated value of *U*.

The Mann–Whitney *U* formula below is given in the Edexcel exam. Note that Table 2 uses capital *N*, A, and B and the Edexcel formula uses small letters.

$$U_A = n_a n_b + \frac{n_a(n_a + 1)}{2} - \Sigma R_A$$

$$U_B = n_a n_b + \frac{n_b(n_b + 1)}{2} - \Sigma R_B$$

You substitute your values in the formula:

$$U_A = 10 \times 14 + \frac{10\,(10 + 1)}{2} - 178.5$$

$$= 140 + 55 - 178.5 = 16.5$$

$$U_B = 10 \times 14 + \frac{14\,(14 + 1)}{2} - 121.5$$

$$= 140 + 105 - 121.5 = 123.5$$

U = the smaller of the two values is the calculated test value = 16.5

Table 2: Final calculation table. Final data table

Column 1	Column 2	Column 3	Column 4
Condition A High arousal	Rank A	Condition B Low arousal	Rank B
7	16	4	4.5
10	23.5	6	12.5
8	18.5	2	1
6	12.5	5	8
5	8	3	2.5
8	18.5	5	8
9	21	6	12.5
7	16	4	4.5
10	23.5	5	8
9	21	7	16
		9	21
		3	2.5
		5	8
		6	12.5
$N_A = 10$ $\Sigma R_A = 178.5$		$N_B = 14$ $\Sigma R_B = 121.5$	

Step 5: Is the result in the right direction?

As the hypothesis is directional you have to check that the result is in the expected direction. The expected direction was that the ratings would be higher in group A than group B. The mean for group A is 17.85 and group B is 8.68, so the result is in the right direction.

Step 6: Find critical value of *U*.

- One-tailed or two-tailed test? The hypothesis is directional so a one-tailed test is used.
- Significance level selected = $p \leq 0.05$ (5% level).
- Degrees of freedom (*df*) or N value: $N_A = 10$ and $N_B = 14$

Now you use the table of critical values (see next page) for a one-tailed test, and locate the row and column that begins with your N values – the critical value of *U* = 41

Table of critical values for the Mann–Whitney test for a one-tailed test ($p \leq 0.05$)

	Critical values at $p = 0.05$ (5%)															
N_A	N_B															
	5	6	7	8	9	10	11	12	13	14	15	16	17	18	19	20
3	1	2	2	3	4	4	5	5	6	7	7	8	9	9	10	11
4	2	3	4	5	6	7	8	9	10	11	12	14	15	16	17	18
5	4	5	6	8	9	11	12	13	15	16	18	19	20	22	23	25
6	5	7	8	10	12	14	16	17	19	21	23	25	26	28	30	32
7	6	8	11	13	15	17	19	21	24	26	28	30	33	35	37	39
8	8	10	13	15	18	20	23	26	28	31	33	36	39	41	44	47
9	9	12	15	18	21	24	27	30	33	36	39	42	45	48	51	54
10	11	14	17	20	24	27	31	34	37	41	44	48	51	55	58	62
11	12	16	19	23	27	31	34	38	42	46	50	54	57	61	65	69
12	13	17	21	26	30	34	38	42	47	51	55	60	64	68	72	77
13	15	19	24	28	33	37	42	47	51	56	61	65	70	75	80	84
14	16	21	26	31	36	41	46	51	56	61	66	71	77	82	87	92
15	18	23	28	33	39	44	50	55	61	66	72	77	83	88	94	100
16	19	25	30	36	42	48	54	60	65	71	77	83	89	95	101	107
17	20	26	33	39	45	51	57	64	70	77	83	89	96	102	109	115
18	22	28	35	41	48	55	61	68	75	82	88	95	102	109	116	123
19	23	30	37	44	51	58	65	72	80	87	94	101	109	116	123	130
20	25	32	39	47	54	62	69	77	84	92	100	107	115	123	130	138

The calculated value of U must be equal to or less than the critical value in this table for significance to be shown.

Table of critical values for the Mann–Whitney test for a two-tailed test ($p \leq 0.05$)

	Critical values at $p = 0.05$ (5%)															
N_A	N_B															
	5	6	7	8	9	10	11	12	13	14	15	16	17	18	19	20
3	0	1	1	2	2	3	3	4	4	5	5	6	6	7	7	8
4	1	2	3	4	4	5	6	7	8	9	10	11	11	12	13	14
5	2	3	5	6	7	8	9	11	12	13	14	15	17	18	19	20
6	3	5	6	8	10	11	13	14	16	17	19	21	22	24	25	27
7	5	6	8	10	12	14	16	18	20	22	24	26	28	30	32	34
8	6	8	10	13	15	17	19	22	24	26	29	31	34	36	38	41
9	7	10	12	15	17	20	23	26	28	31	34	37	39	42	45	48
10	8	11	14	17	20	23	26	29	33	36	39	42	45	48	52	55
11	9	13	16	19	23	26	30	33	37	40	44	47	51	55	58	62
12	11	14	18	22	26	29	33	37	41	45	49	53	57	61	65	69
13	12	16	20	24	28	33	37	41	45	50	54	59	63	67	72	76
14	13	17	22	26	31	36	40	45	50	55	59	64	67	74	78	83
15	14	19	24	29	34	39	44	49	54	59	64	70	75	80	85	90
16	15	21	26	31	37	42	47	53	59	64	70	75	81	86	92	98
17	17	22	28	34	39	45	51	57	63	67	75	81	87	93	99	105
18	18	24	30	36	42	48	55	61	67	74	80	86	93	99	106	112
19	19	25	32	38	45	52	58	65	72	78	85	92	99	106	113	119
20	20	27	34	41	48	55	62	69	76	83	90	98	105	112	119	127

The calculated value of U must be equal to or less than the critical value in this table for significance to be shown.

Step 7: Accept or reject the null hypothesis and draw a conclusion.

For the Mann–Whitney test the calculated (observed) value must be equal to or less than the critical value for your result to be significant. The calculated value (16.5) is less than the critical value (41) so your result is significant at $p \leq 0.05$ (5% significance level).

This means you can reject the null hypothesis and conclude that men find photographs of women more attractive in high arousal states compared to low arousal states ($p \leq 0.05$).

PRACTICE QUESTIONS

These questions are for you to practise what you have learned from sections 3.6 and 3.7

1. A psychology class ask 20 friends to rate out of 10 how much they like psychology after the first month of taking the subject and again 6 months later. The data they collect are shown below:

   After 1 month: 3, 5, 6, 4, 7, 2, 7, 8, 5, 9, 8, 10, 4, 5, 3, 7, 9, 6, 4, 5
   After 6 months: 6, 5, 5, 9, 8, 7, 8, 9, 5, 8, 5, 8, 6, 6, 6, 4, 10, 7, 4, 6

   a. Identify whether the data are related or unrelated.

   b. Explain why the data would be considered ordinal.

   c. Identify an appropriate test to use with the data and justify your choice.

   d. When calculating the Wilcoxon or Mann–Whitney tests you need to rank both sets of data jointly. Perform this ranking task for the data above using the table headings: rating, frequency, ranks, final rank.

2. A research study investigates the effectiveness of a therapy for treating depression. Clients are given a diagnostic test before a three-week programme of treatment and the same test after treatment. The maximum score of 20 indicates a high level of depression. The results for each participant before and after were:

   Client A (15, 16), client B (12, 15), client C (12, 11), client D (19, 13), client E (15, 14), client F (17, 10), client G (10, 11), client H (18, 19), client I (16, 8), client J (15, 15)

   The alternative hypothesis for the study is: Participants show lower scores after treatment than before (note that a reduced score means improvement).

   a. State the level of measurement used in this study.

   b. State whether you would use a Wilcoxon or a Mann–Whitney test to analyse the data and explain your choice.

   c. State a null hypothesis for this study.

   d. Use an appropriate inferential test to determine whether the therapy was effective (at $p \leq 0.05$).

3. A researcher compared the performance of younger and older participants on a task that assessed memory recall. The results from this study were as follows:

   Scores for younger participants (out of 10): 3, 10, 5, 7, 4, 9, 1, 6, 2, 3, 4
   Scores for older participants (out of 10): 6, 3, 7, 5, 2, 4, 8, 9, 7, 8, 4, 1

   The alternative hypothesis for this study is: There is a difference in the accuracy scores of younger and older participants on a task assessing memory recall.

   a. State whether you would use a Wilcoxon or a Mann–Whitney test and explain your choice.

   b. State whether the alternative hypothesis is directional or non-directional.

   c. State a null hypothesis for this study.

   d. Use an appropriate inferential test to determine whether the therapy was effective (at $p \leq 0.05$).

4. For the following examples decide whether a Wilcoxon or a Mann–Whitney test should be used:

   a. The reaction time of cats and dogs are compared to see which group of animals is faster.

   b. An experiment to compare stress levels in doctors and nurses, to see if doctors are more stressed than nurses.

   c. Participants' reaction times are tested before and after drinking coffee.

   d. A study to see if smokers or non-smokers have different attitudes about the environment.

5. Using the table of critical values for the Wilcoxon test in section 3.4, report whether the null hypothesis should be rejected or accepted in the following examples:

   a. 5% significance level, non-directional hypothesis, $N = 15$, no zero values, calculated value of $T = 28$

   b. 5% significance level, non-directional hypothesis, $N = 9$, one zero value, calculated value of $T = 7$

   c. 5% significance level, directional hypothesis, $N = 19$, no zero values, calculated value of $T = 53$

   d. 5% significance level, directional hypothesis, $N = 14$, no zero values, calculated value of $T = 15$

6. Using the tables of critical values for the Mann–Whitney U test in section 3.7, report whether the null hypothesis should be rejected or accepted in the following examples:

   a. 5% significance level, non-directional hypothesis, $N_A = 5$, $N_B = 10$, calculated value of $U = 8$

   b. 5% significance level, non-directional hypothesis, $N_A = 11$, $N_B = 14$, calculated value of $U = 66$

   c. 5% significance level, directional hypothesis, $N_A = 9$, $N_B = 9$, calculated value of $U = 17$

   d. 5% significance level, directional hypothesis, $N_A = 15$, $N_B = 11$, calculated value of $U = 39$

3.8 Related *t*-test

This inferential test is used when looking at **paired** or **related** data (see section 3.6).

This *t*-test can only be used in situations where the **parametric assumptions** are fulfilled.

Parametric data criteria

- The level of measurement must be interval or better (see section 2.6).

- The data must be drawn from a population that can be assumed to have a normal distribution. Note that it is not the sample that must be normally distributed but the population. A normal distribution (see section 2.11) is likely for many physical and psychological characteristics, such as height, shoe size, IQ, and friendliness. Therefore you can justify the use of a parametric test by saying that the characteristic measured is assumed to be normal. However, scores on a test of some psychological variables (such as depression) are likely to be skewed because only a small number of people should get high scores and the bulk will get lower scores (a positive skew).

- The variances of the two samples are not significantly different. The variance is a measure of how spread out a set of data are around the mean. It is the square of the standard deviation. In the case of repeated measures (related samples) any difference in the variances should not distort the result (Coolican, 1996). For independent groups you can check the variances. The variance of one sample should not be more than 4 times the variance of the other.

Where possible, researchers prefer to use parametric inferential tests because these tests are more powerful. Note that parametric tests are quite robust and therefore they can be used unless the parametric assumptions are met quite poorly (or not at all).

WORKED EXAMPLE

To calculate the related *t*-test, let's look at an example. In section 3.6 you analysed a data set about familiar faces using the Wilcoxon test. You will re-analyse this using the related *t*-test to see if the more powerful test shows significance.

Step 1: Reasons for choosing a related *t*-test for this study.

The hypothesis states a *difference* between two sets of data.
The two sets of data are pairs of scores from one person = *related*.
The data are assumed to be *interval* for the purpose of comparing the related *t*-test to the Wilcoxon test, but normally you would have to justify this here.

Step 2: State the hypotheses.

Alternative hypothesis: People rate the more frequently seen face as more likeable than the less frequently seen face (a directional hypothesis).
Null hypothesis: There is no difference in the likeability score for more and less familiar faces.

Step 3: Record data and calculate differences.

The raw data from section 3.6 are shown in columns 1 and 2 in the table on the next page – these are the rating scores for each condition. There are two ratings for each participant.

In order to calculate t you need to work out:

N = the number of participants
d = the differences (d) between columns 1 and 2 (column 3)
Σd = the sum of the differences
Σd^2 = square each difference and add them up (column 4)

Calculation table

Participant	Column 1 Most frequent face	Column 2 Less frequent face	Column 3 d	Column 4 d^2
1	5	2	3	9
2	7	3	4	16
3	3	3	0	0
4	8	4	4	16
5	2	3	-1	1
6	4	5	-1	1
7	5	2	3	9
8	3	4	-1	1
9	6	3	3	9
10	3	5	-2	4
11	7	2	5	25
12	3	4	-1	1
$N = 12$			$\Sigma d = 16$	$\Sigma d^2 = 92$

Step 4: Find the calculated value of t.

The formula for calculating t is:

$$t = \frac{\Sigma d}{\sqrt{((N\Sigma d^2 - (\Sigma d)^2)/(N-1))}}$$

To calculate t you substitute your values (see section 1.9):

$$= \frac{16}{\sqrt{(((12 \times 92) - 16^2)/11)}}$$

$$= 1.82 \text{ (2 d.p.)}$$

Step 5: Is the result in the right direction?

The expected direction was that more frequent face would be more likeable, so the result is in the right direction (the larger value is for positive differences).

Step 6: Find critical value of *t*.

- One-tailed or two-tailed test? The hypothesis is directional so a one-tailed test is used.

- Significance level selected = $p \leq 0.05$ (5% level).

- Degrees of freedom (*df*) = total number of scores. In your case *df* = 12

Now you use the table of critical values (see next page) and locate the column headed 0.05 for a one-tailed test and the row that begins with your *df* value of 12 – the critical value of *t* = 1.782

Table of critical values for the *t*-test

One-tailed	0.05	0.025	0.01	0.005
Two-tailed	0.1	0.05	0.02	0.01
df				
1	6.314	12.706	31.820	63.656
2	2.920	4.303	6.965	9.925
3	2.353	3.182	4.541	5.841
4	2.132	2.776	3.747	4.604
5	2.015	2.571	3.365	4.032
6	1.943	2.447	3.143	3.707
7	1.895	2.365	2.998	3.499
8	1.860	2.306	2.896	3.355
9	1.833	2.262	2.821	3.250
10	1.812	2.228	2.764	3.169
11	1.796	2.201	2.718	3.106
12	1.782	2.179	2.681	3.055
13	1.771	2.160	2.650	3.012
14	1.761	2.145	2.624	2.977
15	1.753	2.131	2.602	2.947
16	1.746	2.120	2.583	2.921
17	1.740	2.110	2.567	2.898
18	1.734	2.101	2.552	2.878
19	1.729	2.093	2.539	2.861
20	1.725	2.086	2.528	2.845
25	1.708	2.060	2.485	2.787
30	1.697	2.042	2.457	2.750

Calculated value of *t* must be equal to or greater than the critical value in this table for significance to be shown.

> **REMEMBER:**
> A useful way to remember whether the calculated (observed) value has to be less than or greater than the critical value is the rule of R.
>
> If the test name has the letter R in it, then it is greater rather than less than.

Step 7: Accept or reject the null hypothesis and draw a conclusion.

For the *t*-test the calculated (observed) value must be equal to or greater than the critical value for your result to be significant. The calculated value (1.82) is equal to or greater than the critical value (1.782) so your result is significant at $p \leq 0.05$ (5% significance level).

This means you can reject the null hypothesis and conclude that people rate the more frequently seen face as more likeable than the less frequently seen face ($p \leq 0.05$).

> **NOTE:**
> Why did you get a different result? The Wilcoxon test only *ranks* the differences and therefore does not detect the very big differences at the higher end of this particular data set. It is a less powerful test.

> **NOTE:**
> Practice questions for this section can be found after section 3.9.

3.9 Unrelated *t*-test

This inferential test is used when looking at **independent** or **unrelated** data (see section 3.7). This test can be used when parametric assumptions are fulfilled (see section 3.8).

WORKED EXAMPLE

To calculate the unrelated *t*-test, let's consider a research study that timed how long it took participants to complete a puzzle. One group were told it was an easy puzzle and the other group were told it was difficult.

Step 1: Reasons for choosing an unrelated *t*-test.

The hypothesis states a *difference* between two sets of data.
The two sets of data are from different people = *unrelated*.
The data are *interval* because they are reaction times, measured on a scale with equal intervals.

Step 2: State the hypotheses.

Alternative hypothesis: People who believe a puzzle is easy complete the puzzle faster than people who believe it is difficult (a directional hypothesis).
Null hypothesis: There is no difference in the speed of completing the puzzle between people who believe it is easy and those who believe it is hard.

Step 3: Record the data.

The raw data are the number of seconds it took to complete the puzzle (see the table on right).

Step 4: Find the calculated value of *t*.

The formula for the unrelated *t*-test is very complex and would not be used in an exam. There are many online sites that will calculate this for you, for example, www.socscistatistics.com/tests/studentttest/
The result of this calculation is $t = 1.882$

Data table showing time to complete puzzle (in seconds)

Group A Difficult belief	Group B Easy belief
74	56
96	34
83	49
80	74
40	61
22	30
97	52
121	83
92	110
73	43
93	
85	
$N_A = 12$	$N_B = 10$
Mean = 79.7	Mean = 59.2

Step 5: Is the result in the right direction?

Yes, the results are in the expected direction. The mean time is faster for the 'easy' expectations group, compared with the 'difficult' expectations group.

Step 6: Find the critical value of *t*.

- One-tailed or two-tailed test? The hypothesis is directional, therefore a one-tailed test is used.
- Significance level selected = $p \leq 0.05$ (5% level).
- Degrees of freedom $(df) = (N_A + N_B) - 2 = 20$
- Now you use the table of critical values for a *t*-test on the previous page, and locate the column headed 0.05 for a one-tailed test and the row that begins with your *df* value of 20. The critical value of $t = 1.725$

Step 7: Accept or reject the null hypothesis and draw a conclusion.

For the t-test the calculated (observed) value must be equal to or greater than the critical value for your result to be significant. The calculated value (1.882) is equal to or greater than the critical value (1.725) so your result is significant at $p \leq 0.05$ (5% significance level).

This means you can reject the null hypothesis and conclude that people who believe a puzzle is easy complete the puzzle in a faster time than people who believe it is difficult (at $p \leq 0.05$)

PRACTICE QUESTIONS

These questions are for you to practise what you have learned from sections 3.8 and 3.9

1. A psychologist was interested to see whether people remembered information better in the morning or in the afternoon. To investigate this she gave some participants a memory test during a morning lesson and some participants took the test during an afternoon lesson. The alternative hypothesis was, 'Students do better on a test given in the morning than in the afternoon.'

   a. Identify whether this is related or unrelated data.

   b. Explain why the data collected would be classed as interval data.

   c. Identify an appropriate parametric test to use with this data and justify your choice.

   d. State whether a one-tailed or two-tailed test is required.

   e. The morning group consisted of 16 students and the afternoon group was 11 students. Identify the critical value of t that would be required for the result to be significant at the 5% level (see the appropriate table of critical values on previous spread).

2. The study above was repeated, but this time each student was tested twice, once in the morning and once in the afternoon.

   Identify an appropriate parametric test to use with this data and justify your choice.

3. A study is conducted comparing each participant's time taken to run 100 metres (measured in seconds) before and after drinking a glass of high energy drink to see if there is a difference. The researchers are not sure whether the drink will speed participants up or actually slow them down.

   a. State whether the hypothesis would be directional or non-directional.

   b. A related t-test is going to be used to analyse the data. The formula is:

   $$t = \frac{\Sigma d}{\sqrt{((N\Sigma d^2 - (\Sigma d)^2/(N-1))}}$$

   Substitute the values below into the formula and, using a calculator, work out the value for t:

   $\Sigma d = 8$, $\Sigma d^2 = 66$, $N = 10$

   c. State whether the null hypothesis can be rejected at a 2% level of significance and explain your conclusion.

4. In the following examples state which inferential test you would choose and justify your choice. The answer could be: the sign test, Wilcoxon test, Mann–Whitney test, related *t*-test, or unrelated *t*-test.

   a. An experiment to compare the scores of older children and younger children on an IQ test.

   b. A study to look at whether identical twin pairs are both colour-blind or not.

   c. Participants are asked to rate pictures of food for attractiveness before eating food and again afterwards to see if food looks more attractive after eating.

   d. A study looks at the effects of sleep on performance by comparing memory test scores in people who slept the previous night either for 8 hours or for 4 hours.

   e. A researcher aims to compare aggressiveness in children whose mothers work and whose mothers don't work. Aggressiveness was determined by asking teachers to rate the children's behaviour.

   f. Participants see a horror film and the next day watch a romantic comedy. After each film their heart rate is recorded to see if it is higher after a horror film.

5. Explain how *df* is calculated for a related and an unrelated *t*-test.

6. A researcher conducts an inferential test and finds a significant result but the data are not in the direction as stated in the hypothesis. Explain if they can accept the alternative hypothesis.

7. Using the tables of critical values for a *t*-test in section 3.8 report whether the null hypothesis should be rejected or accepted in the following examples:

   a. 5% significance level, directional hypothesis, $N = 10$, calculated value of $t = 1.870$

   b. 5% significance level, directional hypothesis, $N = 17$, calculated value of $t = 1.93$

   c. 1% significance level, directional hypothesis, $N = 17$, calculated value of $t = 1.93$

   d. 10% significance level, non-directional hypothesis, $N = 14$, calculated value of $t = 1.33$

3.10 Chi-squared test

Chi-squared (or sometimes written as chi-square) is represented by the Greek letter chi χ. Chi-squared is χ^2

This can be a test of difference or a test of association. You can use it when investigating whether two sets of data are different or to see if there is an association between them. The key features are:

- The test can be used with any kind of data (nominal or better) but it is reduced to categories (nominal) and recorded as frequency data (membership in each category is counted).

- The data can be represented in a contingency table as shown below.

- The data in each category must be independent (the shaded area of the contingency table).

For example, you ask 50 boys and girls to say whether they prefer gold or silver. You then count up how many boys said gold, how many said silver, and the same for the girls. Your results are shown in a 2×2 contingency table:

	Boys	Girls	Totals
Gold	11	15	26
Silver	8	16	24
Totals	19	31	50

The hypothesis could be 'Boys and girls express a *difference* in their liking for gold and silver,' or, 'There is an *association* between gender and liking gold and silver.'

A key point is the independence of the data — you must imagine that each of your 50 people can only be placed in one of the shaded squares.

2×2 means two rows and two columns. You can have any number of rows and columns. For example, you could ask 50 boys and girls whether they prefer gold, silver, or another metal. Then you would have a 2×3 table (2 rows and 3 columns — the figure for rows is always given first).

WORKED EXAMPLE

To calculate chi-squared, let's analyse data from a study that investigated the difference between people who are city- or country-dwellers to see whether they were for or against fox hunting.

Step 1: Reasons for choosing the chi-squared test for this study.

- The hypothesis states a *difference/association* between two variables (a place of residence and an attitude towards fox hunting).

- The data in each cell are *independent*.

- The data are *nominal* because each person belongs to a category.

Step 2: State the hypotheses.

Alternative hypothesis: There is a difference between city- or country-dwellers in terms of whether they agree with fox hunting or not (a non-directional hypothesis). **or** There is an association between where a person lives (the city or country) and their attitude towards fox hunting (a non-directional hypothesis).

Null hypothesis: There is no association/difference between where a person lives (the city or country) and their attitude towards fox hunting.

Step 3: Record the data in a contingency table.

The raw data are: 5 city-dwellers pro-hunting, 10 city-dwellers anti-hunting, 12 country-dwellers pro-hunting, and 9 country-dwellers anti-hunting.

Contingency table

	City	Country	Totals
Pro-fox hunting	5	12	17
Anti-fox hunting	10	9	19
Totals	15	21	36

Step 4: Find the calculated value of χ^2

The expected frequencies are calculated following the steps in the table below.

You then add all the values in the final column to get the calculated value of $\chi^2 = 1.9839$

Calculation table

	row total × column total / total = expected frequency (*E*)	Subtract expected value (*E*) from observed frequency (*O*) (*O* − *E*)	Square previous value (*O* − *E*)^2	Divide previous value by expected frequency (*O* − *E*)^2 / *E*
City pro-hunting	$17 \times \dfrac{15}{36} = 7.08$	$5 - 7.08 = -2.08$	4.3264	0.6110
Country pro-hunting	$17 \times \dfrac{21}{36} = 9.92$	$12 - 9.92 = 2.08$	4.3264	0.4361
City anti-hunting	$19 \times \dfrac{15}{36} = 7.92$	$10 - 7.92 = 2.08$	4.3264	0.5463
Country anti-hunting	$19 \times \dfrac{21}{36} = 11.08$	$9 - 11.08 = -2.08$	4.3264	0.3905

Step 5: Is the result in the right direction?

No direction was stated.

Step 6: Find critical value of χ^2

- One-tailed or two-tailed test? The hypothesis is non-directional so a two-tailed test is used.
- Significance level selected = $p \leq 0.05$ (5% level).
- Degrees of freedom (*df*) = (number of rows − 1) × (number of columns − 1) = 1

Now you use the table of critical values (on the facing page) and locate the column headed 0.05 for a two-tailed test and the row that begins with your *df* value of 1, so the critical value of $\chi^2 = 3.841$

Table of critical values for the Chi-squared (χ^2) test.

One-tailed	0.05	0.025	0.01	0.005
Two-tailed	**0.10**	**0.05**	**0.02**	**0.01**
df				
1	2.706	3.841	5.412	6.635
2	4.605	5.991	7.824	9.21
3	6.251	7.815	9.837	11.345
4	7.779	9.488	11.668	13.277
5	9.236	11.07	13.388	15.086
6	10.645	12.592	15.033	16.812
7	12.017	14.067	16.622	18.475
8	13.362	15.507	188.168	20.09
9	14.684	16.979	19.679	21.666
10	15.987	18.307	21.161	23.209
15	22.307	24.996	28.259	30.578
20	28.412	31.41	35.02	37.566
25	34.382	37.652	41.566	44.314
30	40.256	43.773	47.962	50.892

Calculated value of χ^2 must be equal to or greater than the critical value in this table for significance to be shown.

Step 7: Accept or reject the null hypothesis and draw a conclusion.

For the chi-squared test the calculated (observed) value must be equal to or greater than the critical value for your result to be significant. The calculated value (1.9839) is not equal to or greater than the critical value (3.841) so your result is not significant at $p \leq 0.05$ (5% significance level).

This means you must accept the null hypothesis and conclude that there is no association/difference between where a person lives (the city or country) and their attitude towards fox hunting ($p \leq 0.05$).

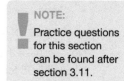

NOTE:
Practice questions for this section can be found after section 3.11.

3.11 Spearman's test

The final two inferential tests are both tests of correlation. You looked at correlational analysis using scatter diagrams (see section 2.13). Scattergrams are a way to perform a simple visual analysis of the relationship between co-variables. In order to determine the exact strength of this relationship an inferential test is needed.

For **non-parametric** data you use Spearman's test (parametric criteria are given in section 3.8).

Two important things to note:

- The figure produced by a correlation test is a number between -1 and $+1$. It is called a **correlation coefficient** and represents the strength of the correlation. -1 is a perfect negative correlation and $+1$ is a perfect positive correlation. The table of critical values (on the facing page) shows that for 15 participants a correlation coefficient around 0.5 is significant.

- To determine the strength of a correlation the sign ($+$ or $-$) is not relevant.

- Correlation coefficients are written without the leading zero.

WORKED EXAMPLE

One way to study genetic influence on intelligence is to test identical twins to see if their intelligence test scores are closely related. If genetic influence is an important factor in intelligence you would expect to find a significant positive correlation.

Step 1: Reasons for choosing the Spearman's test for this study.

The aim states a *correlation* between two sets of data.

The 2 sets of data are pairs of scores from twins = *related*.

At least one of the variables is *ordinal* because IQ is measured using a standardised scale and may not have equal intervals between scores.

Step 2: State the hypotheses.

Alternative hypothesis: There is a positive correlation between the IQ scores of identical twins (a directional hypothesis).

Null hypothesis: There is no correlation between the IQ scores of identical twins.

Step 3: Record the data and calculate ranks and differences.

- Rank data from Twin A from low to high (each data set is ranked separately; see section 2.9).
- Rank data from Twin B from low to high.
- Find the difference (d) between each pair of ranks.
- Square this difference (d^2).
- Add the squared differences (Σd^2).

Calculation table

Twin A	Rank A	Twin B	Rank B	d	d^2
95	5	101	7.5	-2.5	6.25
90	3	86	2	1	1
110	11	108	11	0	0
99	7	111	12	-5	25
120	13	101	7.5	5.5	30.25
85	2	91	4	-2	4
91	4	89	3	1	1
105	10	107	10	0	0
104	9	106	9	0	0
115	12	120	13	-1	1
78	1	82	1	0	0
98	6	98	6	0	0
102	8	95	5	3	9
					$\Sigma d^2 = 77.5$

Step 4: Find the calculated value of *rho*.

The formula is:

$$rho = 1 - \frac{6\Sigma d^2}{N(N^2 - 1)}$$

$$= 1 - \frac{(6 \times 77.5)}{(13 \times (13 \times 13 - 1))}$$

$$= 1 - \frac{465}{2184} = +0.787$$

Step 5: Is the result in the right direction?

Yes, the results were expected to be a positive correlation.

Step 6: Find the critical value of *rho*.

- One-tailed or two-tailed test? The hypothesis is directional so a one-tailed test is used.
- Significance level selected $= p \leq 0.05$ (5% level).
- Degrees of freedom (*df*) or *N* value = 13

Now you use the table of critical values and locate the column headed 0.05 for a one-tailed test and the row that begins with your N value of 13 – the critical value of *rho* = 0.484

Table of critical values for Spearman's *rho*

One-tailed test	0.05	0.025	0.01	0.005
Two-tailed test	0.1	0.05	0.02	0.01
N				
4	1.000			
5	0.900	1.000	1.000	
6	0.829	0.886	0.943	1.000
7	0.714	0.786	0.893	0.929
8	0.643	0.738	0.833	0.887
9	0.600	0.700	0.783	0.833
10	0.564	0.648	0.745	0.794
11	0.536	0.618	0.709	0.755
12	0.503	0.587	0.678	0.727
13	0.484	0.560	0.648	0.703
14	0.464	0.538	0.626	0.679
15	0.443	0.521	0.604	0.654
16	0.429	0.503	0.582	0.635
17	0.414	0.488	0.566	0.618
18	0.401	0.472	0.550	0.600
19	0.391	0.460	0.535	0.584
20	0.380	0.447	0.522	0.570
25	0.337	0.398	0.466	0.511
30	0.306	0.362	0.425	0.467
Calculated value of *rho* must be equal to or greater than the critical value in this table for significance to be shown.				

Step 7: Accept or reject the null hypothesis and draw a conclusion.

For Spearman's test the calculated (observed) value must be equal to or greater than the critical value for your result to be significant. The calculated value (0.787) is equal to or greater than the critical value (0.484) so your result is significant at $p \leq 0.05$ (5% significance level).

This means you can reject the null hypothesis and conclude that there is a positive correlation between the IQ scores of identical twins ($p \leq 0.05$).

REMEMBER:
The sign of the calculated value is not relevant when considering significance.

PRACTICE QUESTIONS

These questions are for you to practise what you have learned from sections 3.10 and 3.11

1. A psychology student surveyed his class of 14 students about their favourite psychologists. He gave them a list of psychologists (Milgram, Bandura, Baddeley) and asked them to select 2 each. Twelve students voted for Milgram, 8 voted for Bandura, and 8 voted for Baddeley.

   Explain why this data could not be analysed using chi-squared.

2. A research study investigates a possible association between handedness and artistic flair (recorded as high or low). The data are analysed using a chi-squared test, with a non-directional hypothesis and a 2×2 contingency table. The observed (calculated) value is 4.23. Identify the level at which this result is significant (using the table of critical values in section 3.10).

3. A research study investigates a possible association between age (over 30 or under 30) and political view (Labour or Conservative voter).

   a. State an alternative non-directional hypothesis for this study.

   b. Sketch a contingency table that would be used to represent this data.

   c. Identify the degrees of freedom.

   d. State the minimum critical value needed for significance to be shown at the 10% level.

4. In another study a set of students ($N = 50$) are asked about their favourite psychological study. This time they could only vote for one psychological study. The researcher found that 18 girls voted for Milgram, 12 boys voted for Milgram, 5 girls voted for Bandura, 6 boys voted for Bandura, 8 girls voted for Baddeley, and 1 boy voted for Baddeley.

   a. Explain why the contingency table for this data would be 2×3

   b. Draw the contingency table for this data.

5. A researcher was studying parenting styles and gave each participant a test to assess (i) what parenting style their parents used (authoritarian or permissive), and (ii) whether their children were high or low in obedience.

   The contingency table below shows the results:

	High obedience	**Low obedience**	**Total**
Authoritarian parent	23	19	42
Permissive parent	15	18	33
Total	38	37	75

   a. State a suitable non-directional alternative hypothesis for this study.

   b. State the null hypothesis.

    c. Explain why you would choose to use chi-squared to analyse this data.

    d. Calculate the value of chi-squared.

    e. Explain what three pieces of information you need to look up the critical value.

    f. State the critical value of chi-squared required.

    g. Explain whether the null hypothesis should be accepted or rejected.

6. A research study examines the hypothesis that couples have similar levels of attractiveness. In other words, if the male partner is judged as very attractive then the female partner also tends to be considered very attractive.

To test this participants are asked to rate photos of male and female partners (couples are not shown together so participants don't know which two people are coupled) and an average (median) rating is given for each photograph. The ratings are shown on the right for each couple.

Couple	Rating for man	Rating for woman
A	9	8
B	5	7
C	5	6
D	4	7
E	8	9
F	7	6
G	8	7
H	3	2
I	6	4

    a. Explain which inferential test would be suitable to analyse the data.

    b. State an alternative directional hypothesis for this study.

    c. Calculate the value of *rho* using the formula below.

$$rho = 1 - \frac{6\Sigma d^2}{N(N^2 - 1)}$$

    d. Explain whether you should accept or reject the null hypothesis.

    e. Imagine that you found that the calculated (observed) value was negative. Explain what this would mean.

    f. If the alternative hypothesis had been non-directional, explain whether the null hypothesis would be accepted with a value of −.371

    g. A friend is suspicious that the calculations were wrong because a positive correlation was expected. Explain how you could sense-check the data.

7. A research study investigates a possible correlation between volume of brain matter and IQ. The hypothesis states that a positive correlation is expected.

    a. Explain why Spearman's test would be suitable to use with this data.

    b. 8 people are assessed, their data ranked, and the sum of their rank differences calculated: $\Sigma d^2 = 31$. Substitute these values in the formula for Spearman's test (see above) and work out the observed (calculated) value. Give your answer to 3 decimal places.

    c. Use the value calculated in part b to determine whether the result is significant.

3.12 Pearson's test

The second inferential test for correlation is a parametric test, meaning you only use it if your data fit parametric criteria (see section 3.8).

WORKED EXAMPLE

Research has suggested that aggression and hot weather are related – as temperatures rise in summer there are more incidents of violence.

Step 1: Reasons for choosing Pearson's test for this study.

- The hypothesis states a *correlation* between two sets of data.
- The two sets of data are data recorded on the same day in one city (air temperature and violent incidents reported to the police) = *related*.
- Temperature is interval because the difference between 22 and 23 is the same as between 30 and 31. Number of incidents is interval because it is a count of frequency.

Step 2: State the hypotheses.

Alternative hypothesis: There is a positive correlation between air temperature and violent incidents reported to the police (a directional hypothesis).

Null hypothesis: There is no correlation between air temperature and violent incidents reported to the police.

Step 3: Record the data.

The data are in the table below:

	June 1	June 8	June 15	June 22	June 29	July 6	July 13	July 20	July 27	Aug 3	Aug 10	Aug 17
Incidents	95	120	100	130	123	142	160	101	125	150	105	128
Temperature in °C	23	25	28	26	22	22	25	27	30	25	31	29

Step 4: Find the calculated value of *r*.

The formula for Pearson's *r* is very complex and would not be used in an exam. There are many online sites that will calculate this for you.

The result for your data is $r = -0.277$

Step 5: Is the result in the right direction?

No, the correlation is a negative one. This means you should stop here because you cannot accept the alternative hypothesis as it is in the wrong direction. However, for the purpose of providing an example, you will continue.

Step 6: Find the critical value of *r*.

- One-tailed or two-tailed test? The hypothesis is directional so a one-tailed test is used.
- Significance level selected = $p \leq 0.05$ (5% level).
- Degrees of freedom (*df*) = $N - 2 = 10$

Now you use the table of critical values and locate the column headed 0.05 for a one-tailed test and the row that begins with your *df* value of 10 – the critical value of $r = 0.497$

Step 7: Accept or reject the null hypothesis and draw a conclusion.

For Pearson's test the calculated (observed) value must be equal to or greater than the critical value for your result to be significant. The calculated value (0.277 — the sign does not matter when considering magnitude) is not equal to or greater than the critical value (0.497) so your result is not significant at $p \leq 0.05$ (5% significance level).

Furthermore the results were not in the expected direction. This means you can accept the null hypothesis and conclude that there is no significant correlation between air temperature and violent incidents reported to the police ($p \leq 0.05$).

Table of critical values for Pearson's r

One-tailed	0.05	0.025	0.01	0.005
Two-tailed	0.1	0.05	0.02	0.01
df (N–2)				
1	0.988	0.997	0.999	0.999
2	0.900	0.950	0.950	0.990
3	0.805	0.878	0.878	0.959
4	0.729	0.811	0.811	0.917
5	0.669	0.755	0.754	0.875
6	0.622	0.707	0.707	0.834
7	0.582	0.666	0.666	0.798
8	0.549	0.632	0.632	0.765
9	0.521	0.602	0.602	0.735
10	0.497	0.576	0.576	0.708
11	0.476	0.553	0.553	0.684
12	0.458	0.532	0.532	0.661
13	0.441	0.514	0.514	0.641
14	0.426	0.497	0.497	0.623
15	0.412	0.482	0.482	0.606
16	0.400	0.468	0.468	0.590
17	0.389	0.456	0.456	0.575
18	0.378	0.444	0.444	0.561
19	0.369	0.433	0.433	0.549
20	0.360	0.423	0.423	0.537

Calculated value of r must be equal to or greater than the critical value in this table for significance to be shown.

PRACTICE QUESTIONS

1. If you are conducting a correlational analysis with data that is interval it is permissible to use a non-parametric test. This would be easier to calculate. However, a researcher opts to use a parametric Pearson's test. Explain why the researcher would prefer to do this.

2. Explain why Pearson's would be a suitable test to use if you are investigating the correlation between a person's height and reaction time.

3. A researcher conducts a series of studies each using a correlational analysis. The data are interval level. For the scenarios below, state whether the null hypothesis would be accepted or rejected.

   a. One-tailed hypothesis, 5% level of significance, $N = 18$, calculated value = +.491

   b. Two-tailed hypothesis, 10% level of significance, $N = 6$, calculated value = −.710

   c. One-tailed hypothesis, 1% level of significance, $N = 11$, calculated value = +.491

1. A researcher wanted to see if people perform better or worse when other people are watching. The researcher devised a standardised procedure to use with the participants and each received exactly the same set of instructions for the study. One group of participants were asked to learn and recite a poem without being watched by others (Group A). A matched group of participants had to do the same task, but in front of an audience of four people (Group B). The researcher recorded how many verbal errors were made.

   a. Identify an inferential test that the researcher could have used to analyse the results. Give two reasons why this test was appropriate to use for this study.

   b. The researcher used a non-directional hypothesis. Explain the difference between a directional and a non-directional hypothesis.

   c. The researcher found that the difference between the measurements in the control and experimental condition was significant at $p \leq 0.05$. Explain what is meant by 'significant at $p \leq 0.05$'.

2. A researcher wanted to test the hypothesis that intelligence and talkativeness are positively correlated. To that end, 20 participants were given an IQ test. The researcher then interviewed the same 20 participants, and the amount of time they spent talking in the 30 minute interview was recorded.

   a. Is the researcher's hypothesis directional or non-directional? Explain your answer.

   b. Name an inferential test the researcher could have used to test the hypothesis. Give two reasons for your choice.

   c. The researcher found their results were significant at the 5% significance level. What is the likelihood that they made a Type 1 error?

3. A researcher wanted to investigate the effect of meaningfulness on the retention of verbal material. Seven participants were assigned to a condition in which they were shown a list of 10 'meaningful' words (Condition A). Seven different participants were shown a list of 10 'nonsense' words (Condition B). The number of words correctly recalled was the dependent variable. The researcher's hypothesis stated that more material would be recalled in Condition A than in Condition B. Unfortunately, the data from 1 participant in Condition A was misplaced, leaving only six participants in that condition.

   a. The researcher decided to use a Mann–Whitney test to analyse the results. Give two reasons why she chose to use this test.

   b. Give one reason why Spearman's *rho* test would have been inappropriate to analyse the data in this study.

   c. The researcher found a value of $U = 3$. Using the appropriate table (see facing page), explain whether the researcher's findings are significant at the 0.05 significance level for a directional hypothesis.

4. A teacher decided to test the hypothesis that people will be less able to solve mathematical problems when listening to Taylor Swift songs through headphones than when listening to Pink Floyd songs through headphones. The teacher used 20 participants and asked them to solve a series of mathematical problems while listening to Taylor Swift. She then asked them to solve a similar series of mathematical problems while listening to Pink Floyd. 13 participants solved more problems when listening to Pink Floyd, whereas only 5 solved more problems when listening to Taylor Swift. Two participants scored the same amount in the two conditions.

   a. Is the teacher's hypothesis directional or non-directional?

   b. The teacher used the sign test to analyse her data because it can be calculated quickly. Give two other reasons why this test was suitable for her data.

Table of critical values for the two-tailed Mann–Whitney test ($p \leq 0.05$)

N_A	N_B															
	5	6	7	8	9	10	11	12	13	14	15	16	17	18	19	20
3	0	1	1	2	2	3	3	4	4	5	5	6	6	7	7	8
4	1	2	3	4	4	5	6	7	8	9	10	11	11	12	13	14
5	2	3	5	6	7	8	9	11	12	13	14	15	17	18	19	20
6	3	5	6	8	10	11	13	14	16	17	19	21	22	24	25	27
7	5	6	8	10	12	14	16	18	20	22	24	26	28	30	32	34
8	6	8	10	13	15	17	19	22	24	26	29	31	34	36	38	41
9	7	10	12	15	17	20	23	26	28	31	34	37	39	42	45	48
10	8	11	14	17	20	23	26	29	33	36	39	42	45	48	52	55
11	9	13	16	19	23	26	30	33	37	40	44	47	51	55	58	62
12	11	14	18	22	26	29	33	37	41	45	49	53	57	61	65	69
13	12	16	20	24	28	33	37	41	45	50	54	59	63	67	72	76
14	13	17	22	26	31	36	40	45	50	55	59	64	67	74	78	83
15	14	19	24	29	34	39	44	49	54	59	64	70	75	80	85	90
16	15	21	26	31	37	42	47	53	59	64	70	75	81	86	92	98
17	17	22	28	34	39	45	51	57	63	67	75	81	87	93	99	105
18	18	24	30	36	42	48	55	61	67	74	80	86	93	99	106	112
19	19	25	32	38	45	52	58	65	72	78	85	92	99	106	113	119
20	20	27	34	41	48	55	62	69	76	83	90	98	105	112	119	127

The calculated value of U must be equal to or less than the critical value in this table for significance to be shown.

c. Calculate the value of S. Explain how you arrived at this figure.

d. Using the table on the right, explain whether or not the value of S that you calculated in part c is significant at the 0.05 significance level.

5. A researcher was interested in how people describe themselves in the 'lonely hearts' columns of a national newspaper. He predicted that male advertisers would be more likely than female advertisers to offer financial resources, whilst female advertisers would be more likely than male advertisers to ask for financial support. After reading through various lonely hearts columns in several newspapers and magazines, the researcher found 5 instances of male advertisers asking for financial resources in a partner and 59 instances of female advertisers doing the same thing. The researcher also found 38 instances of male advertisers offering financial resources to a partner and 26 instances of female advertisers doing the same thing.

a. Which level of measurement was used in this study?

b. Identify which inferential test would have been used to analyse the data in this study.

c. The researcher used the 1% significance level. What is the probability of a Type 1 error occurring?

Table of critical values of S

	p value			
One-tailed test	0.05	0.025	0.01	0.005
Two-tailed test	0.1	0.05	0.02	0.01
df				
5	0			
6	0	0		
7	0	0	0	
8	1	0	0	
9	1	1	0	
10	1	1	0	0
11	2	1	1	0
12	2	2	1	0
13	3	2	1	0
14	3	2	2	1
15	3	3	2	1
16	4	3	2	1
17	4	4	3	1
18	5	4	3	2
19	5	4	4	2
20	5	5	4	2

The calculated value of S must be equal to or less than the critical value in this table for significance to be shown.

References

Bandura, A., Ross, D. and Ross, S.A. (1961) Transmission of aggression through imitation of aggressive models. *Journal of Abnormal and Social Psychology, 63*, 575–582.

Baron-Cohen, S., Jolliffe, T., Mortimore, C. and Robertson, M. (1997) Another advanced test of theory of mind: evidence from very high functioning adults with autism or Asperger Syndrome. *Journal of Child Psychology and Psychiatry*, 38, 813–822.

Coolican, H. (1996) *Introduction to research methods and statistics in Psychology*. London: Hodder and Stoughton.

Dement, W.C. and Kleitman, N. (1957) The relation of eye movements during sleep to dream activity: An objective method for the study of dreaming. *Journal of Experimental Psychology*, 53, 339–346.

Gottesman, I.I., Laursen, T.M., Bertelsen, A. and Mortensen, P.B. (2010) Severe mental disorders in offspring with 2 psychiatrically ill parents. *Archives of General Psychiatry*, 67, 252–257.

Haslam, S.A., Reicher, S.D. and Birney, M. E. (2014) Nothing by mere authority: Evidence that in an experimental analogue of the Milgram paradigm participants are motivated not by orders but by appeals to science. *Journal of Social Issues*, 70 (3), 473–488.

Loftus, E.F. and Palmer, J.C. (1974) Reconstruction of automobile destruction: An example of the interaction between language and memory. *Journal of Verbal Learning and Verbal Behavior*, 13, 585–589.

Milgram, S. (1963) Behavioural study of obedience. *Journal of Abnormal and Social Psychology, 67, 371*–378.

Piliavin, I.M., Rodin, J. and Piliavin, J.A. (1969) Good Samaritanism: An underground phenomenon. *Journal of Personality and Social Psychology, 13*, 1200–2013.

Raine, A., Buchsbaum, M. and LaCasse, L. (1997) Brain abnormalities in murderers indicated by positron emission tomography. *Biological Psychiatry*, 42 (6), 495–508.

Roediger, H.L. and Karpicke, J.D. (2006) Test-enhanced learning: Taking memory tests improves long-term retention. *Psychological Science*, 17, 249–255.

Rosenhan, D.L. (1973) On being sane in insane places. *Science*, 179, 250–258.

Simons, D.J. and Chabris, C.F. (1999) Gorillas in our midst: sustained inattentional blindness for dynamic events. *Perception*, 28, 1059–1074.

Van IJzendoorn, M.H. and Kroonenberg, P.M. (1988) Cross-cultural patterns of attachment: A meta-analysis of the Strange Situation. *Child Development*, 59, 147–156.

ANSWERS

1.1 Fractions

1. **a.** Total is 20, so red fraction is $\frac{4}{20}$

   b. $\frac{1}{5}$

   c. Divide each by 28 = $\frac{1}{4}$

   d. $\frac{48}{8}$ = 6

   e. $\frac{48}{8} \times 3$ = 18

2. **a.** $\frac{50}{105}$ (divide each by 5) = $\frac{10}{21}$

   b. $\frac{80}{105}$ (divide each by 5) = $\frac{16}{21}$

   c. $\frac{105}{3}$ = 35

   d. $\frac{105}{5} \times 2$ = 42

3. **a.** $\frac{336}{1008}$ (divide each by 336) = $\frac{1}{3}$

   b. $\frac{1}{3}$ of 100 ($\frac{1}{3}$ to get 0.33, then $\times$ 100)
   = 33 people (the answer is 33 and a third

   but you can't have a third of a person, so you
   would round down)

   c. $\frac{126}{1008}$ (divide each by 126) = $\frac{1}{8}$

   d. $\frac{1}{8}$ of 100 = 13 people (the answer is 12 and
   a half, but this time you should round up)

1.2 Decimals and decimal places

1. **a.** $\frac{22}{100} \times 44$ or 0.22 $\times$ 44

   b. $\frac{3}{4}$ = 0.75

   c. $\frac{5}{6}$ = 0.833333 = 0.8 (1 d.p.)

   d. $\frac{6}{7}$ = 0.85714 = 0.86 (2 d.p.)

   e. $\frac{3}{13}$ = 0.230769 = 0.2 (1 d.p.)

2. **a.** Total number of participants is 77 + 31
   = 108. Females = $\frac{77}{108}$ = 0.713 (3 d.p.)

   b. Males = $\frac{31}{108}$ = 0.287 (3 d.p.)

   c. 4.0 (removing '29' does not require rounding
   up but you must include the 0 to show you
   have one decimal place)

3. **a.** 12.14

   b. 4.92

   c. 0.40

   d. 30.00

   e. 26.11

1.3 Percentages

1. **a.** Total is 20, so red fraction is $\frac{4}{20}$ = 0.2, 0.2 $\times$
   100 = 20%

   b. 50% = $\frac{1}{2}$, and half of 16 is 8 red balls

   c. $\frac{17}{2}$ = 8.5, rounded up to 9 red balls

   d. Move decimal point two places to right
   = 15.8%

   e. Move decimal point two places to right = 20%

   f. Divide by 100, move decimal point two
   places to left = 0.02

   g. $\frac{1}{8}$ = 0.125 $\times$ 100 = 12.5%

   h. $\frac{3}{8}$ = 0.375 $\times$ 100 = 37.5%

   i. $\frac{64}{100}$, in its simplest form = $\frac{16}{25}$

2. **a.** $\frac{10}{23}$

   b. Texting = $\frac{10}{23}$ = 43%;

   answering questions = $\frac{18}{23}$ = 78%

   out of seats = $\frac{5}{23}$ = 22%

   talking to someone else = $\frac{20}{23}$ = 87%

   c. 80% of 23 = 0.80 $\times$ 23 = 18.4 = 18
   students

3. a. $DN = \dfrac{17}{26} = 65.4\%$

$IR = \dfrac{26}{34} = 76.5\%$

$KC = \dfrac{36}{40} = 90\%$

b. For example: There are quite large individual differences, which suggests other factors may affect the number of dreams and not just the REM stage.

There seems to be an association between dreaming and REM sleep which in turn suggests that some aspect of REM sleep is involved in dreaming.

c. It makes comparisons possible – a fraction for DN of $\dfrac{17}{26}$ is difficult to compare with that for KC $\left(\dfrac{36}{40}\right)$. It is much easier to compare 65.4% and 90%

1.4 Ratios

1. a. 1:6

b. Ratio of red to blue balls is 4:16, simplified to 1:4

c. 4:20, simplified to 1:5

d. 16:20, simplified to 4:5

e. The whole = 3 + 4 = 7, $\dfrac{21}{7} = 3$, therefore 1 child gets $3 \times 3 = 9$ mini-chocolate bars and the other gets $3 \times 4 = 12$ mini-chocolate bars.

2. a. 50:30, simplified to 5:3

b. The 'whole' = 5 + 3 = 8, $\dfrac{120}{8} = 15 \times 5$ (the 'part' for girls) = 75 girls should be selected.

c. $\dfrac{39}{8} = 4.875 \times 5 = 24.375$

You can't recruit 24.375 girls, so this is rounded down to 24 girls.

3. a. 17:6

b. 17:23

c. $\dfrac{6}{23}$

d. $\dfrac{6}{23} \times 100 = 26.087$
$= 26.1\%$ (1 d.p.)

1.5 Standard form

1. a. 1×10^2

b. 3×10^2

c. 4×10^6

d. 3.6×10^3 or 4×10^3

e. 6.672×10^3 or 6.67×10^3 or 6.7×10^3 or 7×10^3

f. 3.9×10^{-5} or 4×10^{-5}

g. 1×10^{-3}

2. a. 1 000 000

b. 56 300

c. 2 900 000 000

d. 0.000 01

e. 0.000 563

f. 0.000 000 002 9

1.6 Estimate results

1. a. 6000 is best. 5500 is closer but would make calculations more difficult.

b. 500 is best. If 400 you would underestimate and if 450 or 460 you may have difficulty doing the calculation in your head.

2. a. i. $200 \times 150 = 30\,000$ (2×15 and add 3 zeros)

ii. 38 718

b. i. $20 \times 400\,000 = 8\,000\,000$ (2×4 and add 6 zeros)

ii. 7 252 192

c. i. $\dfrac{200}{10} = 20$

ii. 29.875

d. i. $\dfrac{3}{16}$ is close to $\dfrac{4}{16} = \dfrac{1}{4}$ of 40 000 = 10 000

ii. 8321.4375

e. i. $\dfrac{5}{8}$ is about $\dfrac{1}{2}, \dfrac{1}{2}$ of 300 = 150

ii. 166.875

f. i. 78% is about 75% = $\dfrac{3}{4}$ of 4000 = 3000 or 78% is about $0.7 \times 4000 = 2800$ (round 0.78 down to compensate for rounding 3527 up).

ii. 2751.06

1.7 Order of magnitude calculations

1. **a.** Compare the powers in 1.22×10^5 and $4.2 \times 10^4 = 5 - 4$, so the first number is 1 order of magnitude bigger (i.e., 10 times bigger).

    b. Compare 4.5×10^6 and $1.8 \times 10^3 = 6 - 3$, so the first number is 3 orders of magnitude bigger (i.e. 1000 times bigger).

    c. Compare 2×10^{-2} and $5.2 \times 10^{-4} = -2 - (-4)$, so the second number is 2 orders of magnitude smaller (i.e. 100 times smaller).

    d. Compare 4.5×10^6 and $2 \times 10^{-2} = 6 - (-2)$, so the first number is 8 orders of magnitude bigger (i.e. 100 000 000 times bigger).

1.8 Significant figures

1. **a.** 105
    b. 104.57
    c. 2400
    d. 2433
    e. 2432.90
    f. 6.1
    g. 6.15
    h. 0.0051
    i. 0.005 138
    j. 0.000 47
    k. 0.070
    l. 0.0702

1.9 Algebra

1. **a.** $x = \dfrac{37.5}{100} \times 69 = 25.875 = 26$ to the nearest whole number

    b. $S = \sqrt{\dfrac{194}{(15 - 1)}} = \sqrt{13.8} = 3.72$ (2 d.p.)

    c. $U_a = 12 \times 10 + \dfrac{(12 \times 13)}{2} - 180 = 18$

Summary questions for Section 1

1. **a.** 38:65

    b. You can estimate that the ratio is about 40:60 which is 1:3, or you might estimate it as 40:70, which is closer to 1:2

    c. $\dfrac{62}{65}$

    d. $\dfrac{62}{65} \times 100 = 95.38 462 = 95.4\%$

    e. $\dfrac{19}{38} = \dfrac{1}{2}$

    f. $\dfrac{19}{38} \times 100 = 50$ to one decimal place $= 50.0\%$

    g. $\dfrac{87}{100} \times 65 = 56.55$, which means that the number of people would be 56 (you cannot round up because if 57 people helped the answer would be $\dfrac{57}{65} \times 100 = 87.769 = 88\%$)

    h. The number of drunk trials was 38, which is about 40 out of about 100 trials in total = about 40%

2. **a.** 86 000 000 000, 260 000 000 000, and 4 700 000 000

    b. 8.6×10^{10}, 2.6×10^{11}, and 4.7×10^9 or 9×10^{10}, 23×10^{11}, and 5×10^9

    c. Compare 8.6×10^{10} and $2.6 \times 10^{11} = 10 - 11$, so an African elephant has about 10 times as many neurons.

    d. 8.6×10^{10} and $4.7 \times 10^9 = 10 - 9$, so a human has about 10 times more neurons than a lion

    e. 1400 cubic centimetres.

3. **a.** one

    b. 41 (smashed), 34 (hit), and 32 (contacted)

    c. There were 9 participants in each condition, $\dfrac{9}{45} \times 100 = 20\%$

4. **a.** $\dfrac{65}{100} \times 40 = 26$

    b. 12.5%

    c. 5:26

    d. 5:26 is approximately 5:25 or 1:5

    e. $\dfrac{65}{100} \times 86 = 55.9$, therefore you would expect 56 to be fully obedient

5. **a.** 32.41 and 20.99

    b. 32.4 and 21.0

    c. $\dfrac{32.4082}{50} \times 100 = 64.8164 = 65\%$

    $\dfrac{20.9901}{50} \times 100 = 41.9802 = 42\%$

    d. 32.4082 is about 32, and $\dfrac{32}{50}$ as a percentage is 64%

    20.9901 is about 21, and $\dfrac{21}{50}$ as a percentage is 42%

    e. $\dfrac{48}{100} \times 50 = 24$ words

6. **a.** $\dfrac{1}{3}$

   b. $\dfrac{72}{3} = 24$

   c. $\dfrac{70}{100} \times (24 + 24) = 33.6$, therefore about 34 of the children

7. **a.** Roughly 2 out of 40, which is $\dfrac{1}{20} = 5\%$

   b. 4.9%

   c. 41

   d. $\dfrac{23}{41} \times 100 = 56.0976 = 56.1\%$

8. **a.** $\dfrac{168}{8} = 21$ cm

   b. Multiplying by 10 000 means moving the decimal point 4 places to the right, so 100 centimetres (or 1 metre).

   c. $1 \times 10^{-7} = 0.000\,000\,1 \times 30 = 0.000\,003$ $= 3 \times 10^{-6}$

9. **a.** $(3 - 1) \times (4 - 1) = 2 \times 3 = 6\ df$

   b. $6df = (x - 1) \times (2 - 1)$. So x = 7

   c. $8df = (3 - 1) \times (x - 1)$. So x = 5

2.1 Quantitative and qualitative data

1. **a.** quantitative

   b. quantitative

   c. Initially qualitative, but can be turned into quantitative by counting the yeses and noes.

   d. qualitative

   e. quantitative

2. **a.** The data produced must be numerical, for example, 'Rate how much your children like the following foods on a scale of 1 to 5, where 5 is strong liking and 1 is strong disliking'.

   b. The data produced must be non-numerical, for example, 'List five foods your child likes to eat'.

3. **a.** A list of behaviours might be drawn up (e.g., changing sitting position, fidgeting) and then each of these counted.

   b. People might be interviewed after seeing the doctor and asked to describe how they felt while waiting.

   c. It will be easier to analyse and draw conclusions from quantitative data compared with qualitative data because you can display it on a graph and also calculate the mean values.

4. **a.** quantitative

   b. It could be regarded as quantitative because because the end data was numerical (numbers in each category). However, the categories are qualitative descriptions (imitative, etc.). So the data recorded for each participant was qualitative.

   c. quantitative

2.2 Primary and secondary data

1. **a.** secondary data

   b. primary data

   c. secondary data

   d. primary data

   e. secondary data

2. **a.** The data was collected by the government for their own purposes of record keeping and the researchers made use of this existing data.

   b. It would have been impossible to gain access to such a large data set except by using official records, so secondary data allows you to investigate something you wouldn't otherwise be able to do.

   c. There are likely to be some people with schizophrenia who are not admitted as psychiatric patients which means the correlations might be an underestimate.

3. **a.** The researchers asked participants to recognise the faces in photographs – this is the data collected, not the photographs themselves.

   b. It would be difficult to find this data from another source as it was very specific to the study's aims.

2.3 Sampling participants and 2.4 Sampling observations

1. **a.** self-selecting/volunteer

   b. systematic

   c. opportunity

   d. stratified

   e. random

   f. opportunity

   g. self-selecting/volunteer

   h. systematic

2. **a.** Event sampling would be best because it might be hard to determine when each activity begins and ends. It is therefore better to note when each child engages in one event.

   b. Time sampling would be best because you are interested in the frequency of the activity and therefore need to look every 15 seconds to note what the students are doing. A picture of frequency can then be constructed.

3. **a.** The psychologist could go to a school and use a group of students from one class.

   b. You could get people who were highly motivated to take part and would stick with the study, and could target particular populations such as science or arts students to see if there was a difference.

   c. This would mean the proportions of different groups might better represent the population being studied and therefore the results would generalise better. For example, the psychologist might prefer to select people from different ethnic groups to represent the target population.

4. **a.** The researcher might advertise for people with pets to take part − a self-selected (volunteer) sample.

   b. The researcher would tally each time the animal started any one of these behaviours.

   c. Time sampling would provide a more informative record because it would reflect the frequency of the behaviours. For example, it would be more useful to know that the animals slept for 4 hours and groomed for 2 hours each day rather than knowing the animal slept twice and groomed 15 times.

2.5 Measures of central tendency

1. **a.** The numbers roughly add up to $105 = \dfrac{105}{10} = 10.5$

   b. Total $= \dfrac{114}{11} = 10.36$ (2 d.p.)

   c. In order the numbers are: 2, 4, 6, 7, 9, 10, 12, 12, 15, 18, 19, the middle (median) is 10

   d. 12, as the only number that appears twice

   e. The mean is best as there aren't any outliers and it provides the most detailed average.

2. **a.** For example, the estimate might be 430, mean = 43 for 10 values

   b. 43.3

   c. In order the numbers are 4, 13, 25, 31, 34, 41, 48, 54, 68, 115. So the median lies between 34 and 41 = 37.5

   d. There is no mode.

   e. Median is better because of the outlying value at the upper end.

2.6 Levels of measurement

1. For example, analysing the grades a set of students got. Count up all of the Grade As, Bs, etc., and the grade with the highest count is the mode.

2. IQ test items may not be equal in terms of difficulty. Also actual score given uses standardised scores that are not a simple reflection of the number of actual items that were scored correctly.

3. Ordinal data is in order but the difference between each item on a scale is not necessarily equal whereas for interval level data the data have equal intervals.

2.7 Measures of dispersion: range and standard deviation

1. **a.** Mean = 12.29, range = 20 + 1 = 21

   b. Mean = 12.75, range = 22 + 1 = 23

2. Set B because, even though the range is similar to Set A, the values are more spread out/dispersed from the mean.

3. $\sqrt{\dfrac{62}{8}} = 2.8$ (1 d.p.), if n is used, the answer is 2.6 (1 d.p.)

4. Therapy A has a bigger improvement rate and therefore seems to be slightly more effective.

   Therapy B has a smaller standard deviation, which suggests it is more consistent and therefore probably the best to use because the mean isn't that much lower than for Therapy A.

Halfway summary questions

1. **a.** Opportunity, because they just used whoever got on the subway train.

   b. It was collected by the researchers for the purpose of this study.

   c. The number of people who helped a confederate.

d. The recorded conversations of some of the passengers.

e. The quantitative data is useful because you can easily see what kind of person was helped more. The qualitative data gives insight into why people do or don't help, such as 'You feel so bad when you don't know what to do'. Together they provide different kinds of information that give a fuller picture of helping behaviours.

f. It is interval/ratio level data as it has a true zero value and six people is twice the value of three, making the intervals equal.

2. a. Self-selected/volunteer, because the people made the decision themselves about whether to take part when they read the newspaper ad.

b. People who volunteer tend to be more motivated to be helpful and this might lead them, for example, to try to produce the data that Milgram wanted. They may therefore have been more obedient than people normally would be.

c. It enabled Milgram to get a good cross-section of the local population because lots of different people saw his adverts.

d. Haslam *et al.* are using data collected by someone else for different research aims – to show that participants did not obey when the order required blind obedience.

e. It is nominal data, because it describes what the people did. There would be categories – obeyed and did not obey – and then you count the frequency of the people in each category.

3. a. opportunity

b. Event sampling because observations were only recorded for specific events.

c. The data would be nominal because you would have descriptions of what the nurses, doctors, and the real patients said/did, and which could be put in different categories.

d. 52 − 7 + 1 = 46 days.

e. 1 is added because the pseudopatient may not have been there for a whole day.

f. There might be one pseudopatient who stayed for a particularly short or long time and this would bias the mean (an outlier). In this case the median might therefore better represent the typical value.

4. a. The larger standard deviation shows that there was a bigger variation in the total pass counts, which suggests that some participants were very inaccurate.

b. If a participant's scores were widely different from the mean scores it suggests they were not really paying attention and that their data was meaningless.

c. The answers would have been in words even if it was just a 'yes' answer.

5. a. The method is self-selected (volunteer) sampling. Advertising in a magazine read by people with autism or interested in autism is more likely to target people with autism than an advert in a local newspaper.

b. The researchers would have the list of names of people on the panel. The names could then be placed in a hat and the appropriate number of names drawn out. Those people would be asked if they will take part.

c. It could be snowball sampling by identifying one person with Tourette syndrome and asking them for other contact names. It could also be volunteer (self-selected) sampling by placing an ad at a clinic attended by people with Tourette syndrome.

d. The data was both quantitative (because it was scores) and primary (because it was collected for the purpose of this study).

e. The scores could be regarded as ordinal because some of the eyes might have been more difficult to identify than others. Therefore getting 8 right might not represent doing twice as good as getting 4 right.

2.8 Frequency tables and 2.9 Ranking data

1. a.

Name	Tally	Frequency
Asch	I	1
Bandura	I	1
Freud	II	2
Loftus	II	2
Milgram	IIII	4
Zimbardo	III	3

b. Milgram

c. nominal

d. The data at the beginning is in words. The frequency was then counted, which is quantitative.

2. a.

Score on memory test	Tally	Frequency	Cumulative frequency
3	II	2	2
4			
5	II	2	4
6	I	1	5
7	IIII I	6	11
8			
9	IIII	5	16
10	II	2	18
11	III	3	21
12	II	2	23
13	I	1	24
14	IIII	4	28
15	III	3	31
16	I	1	32
17			
18			
19	I	1	33

b.

Score on memory test	Tally	Ranks	Final ranks
3	II	1, 2	1.5
4			
5	II	3, 4	3.5
6	I	5	5
7	IIII I	6, 7, 8, 9, 10, 11	8.5
8			
9	IIII	12, 13, 14, 15, 16	14
10	II	17, 18	17.5
11	III	19, 20, 21	20
12	II	22, 23	22.5
13	I	24	24
14	IIII	25, 26, 27, 28	26.5
15	III	29, 30, 31	30
16	I	32	32
17			
18			
19	I	33	33

c. Range = 19 − 3 = 16, mode = 7, median = 17th value = 10

d.

Score on Maths test	Tally	Frequency
1-4	II	2
5-8	IIII IIII	9
9-12	IIII IIII II	12
13-16	IIII IIII	9
17-20	I	1

2.10 Frequency diagrams: bar charts and histograms

1.

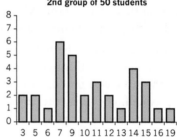

Bar chart showing test scores for 2nd group of 50 students

2. In a histogram (but not a bar chart) you cannot use nominal data.

There must be no spaces between the bars in a histogram, but there must be in a bar chart.

The area of the bars is proportional to the frequency (the axes must start from zero) in a histogram, but not in a bar chart.

3. Add a title ('Bar chart showing students' favourite colours').

Label the y-axis ('Colour choice').

Label the x-axis ('Frequency').

The bars should have spaces between them.

4. a. Some words have a greater effect on the estimate of speed (the answer 'smashed led to the highest estimate' is a statement of fact, not a conclusion).

b. The data on the y-axis are nominal, in other words not continuous. Also frequency is not being measured.

5. a. When tested after 5 minutes the study group did better, *suggesting* in the short-term this is a more effective technique.

However, over time the study-test group did best, *showing* that it is a better long-term strategy.

b. Because the y-axis does not show frequencies and the x-axis is not continuous data.

2.11 Normal distribution

1. It is bell-shaped.

   The mean, median, and mode are all at (approximately) the same point.

   The spread of the data is fixed so, for example, about 68% of the scores fall within one standard deviation of the mean.

2. **a.** Approximately 68%.

   b. Approximately 16% (34% have an IQ of 85–100 points).

3. **a.** 15.87%

   b. 2.28%

   c. 68.26%

2.12 Skewed distributions

1. **a.** In a normal distribution the mean, median, and mode would have very similar values, whereas in a skewed distribution the mean and median are either to the left or to the right of the mode.

   b.
   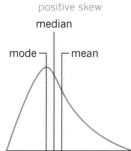

   c. The tail of a negative skew goes in a negative (left) direction. The median is to the left of the mode, and the mean is to the left of the median.

2. You would expect a negative skew because these people are likely to get high scores, so the mode will be to the right with the tail going to the left.

3. **a.** A negatively skewed distribution.

   b. The most likely explanation is that the majority of the students had revised for the test and clearly understood the topic they had been taught. Alternatively or additionally, the test was too easy for the majority of students.

   c. She could make some questions on the test harder. This would mean that fewer students are likely to get a very high mark.

2.13 Scatter diagrams

1. **a.**

   b. A perfect positive correlation, all dots in a straight line.

   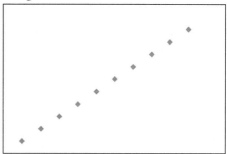

   c.
   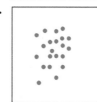

   d.

2. **a.** Moderate negative correlation, dots loosely arranged from top left to bottom right.

   b. Moderate-strong negative correlation, dots arranged fairly closley in a straight line from top left to bottom right.

   c. Moderate-strong positive correlation, dots arranged fairly closely in a straight line from bottom left to top right.

   d. No correlation, dots form no pattern.

3. **a.**
   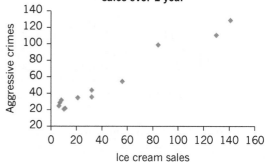
   Scatter diagram showing Mr Gelato's ice cream sales over 1 year

b. There is a strong positive correlation (it is actually +0.98).

c. Hotter temperatures in the summer might increase both ice cream sales and aggression.

Summary questions for Section 2

1. a.

Score	Tally	Total frequency	Ranks
9	I	1	1
10	I	1	2
11	I	1	3
12	II	2	4, 5
13	I	1	6
14			
15	II	2	7, 8
16	III	3	9, 10, 11
17	II	2	12, 13
18	IIII	4	14, 15, 16, 17
19	III	3	18, 19, 20
20	III	3	21, 22, 23
21	II	2	24, 25
22	II	2	26, 27
23			
24	II	2	28, 29
25	I	1	30
26	I	1	31
27			
28	I	1	32

b. Median = 18

c. They should both be 18

d. Mean = 18.16 (2 d.p.), median = 18, mode = 18

e. The data (test scores) are continuous and frequency is shown.

f.

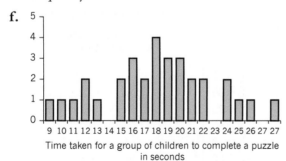

Time taken for a group of children to complete a puzzle in seconds

g. It is roughly bell-shaped and the mean, median, and mode are around the same point.

h.

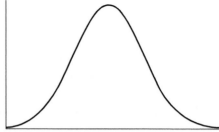

i.

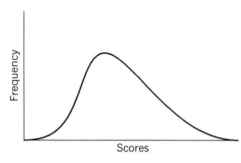

j. It would be smaller because, in a positively skewed distribution, the median is smaller than the mean.

k. Data that is positively skewed has most scores at the lower end. In other words, the second group of students responded more quickly than than the third group of students who showed a negatively skewed distribution.

2. a. positive correlation

b. The easiest would be an opportunity sample (e.g., ask other people in a university department) or a volunteer sample (advertise for people to rate the pictures).

c.

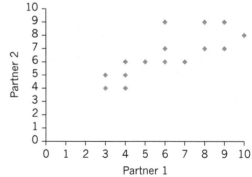

d. It is primary data, and quantitative.

e. The scatter diagram shows a strong positive correlation so the correlation coefficient should be more than +.50 (it is actually +.77).

f. It might be a zero correlation (no relationship in terms of attractiveness) or might be a negative correlation (partners tend to be opposites in terms of attractiveness).

g. In a positive correlation both scores increase together – as in the case of two partners having a similar rating. In a negative correlation as one score increases the other decreases – for example, attractive partners having less attractive partners.

3. a. Primary data = measuring taxi drivers' brain/hippocampus, secondary data = data from control participants because their records were pre-existing.

b. For example, the conclusion you might draw is that this increasing size of the right hippocampus is related to the demands of being a taxi driver (using your memory to remember places) and not an innate characteristic.

c. The bars are not touching, the categories are nominal and not continuous, and the y-axis is not frequency data.

4. a. It means that people who are more obedient are also more likely to have an authoritarian personality.

b. Correlation does not demonstrate a cause. Other variables may be responsible for the link, for example, whatever causes an authoritarian personality (e.g., harsh parenting) may also mean that someone is more scared about not obeying.

3.1 Simple probability

1. $\dfrac{6}{30}$ or $\dfrac{1}{5}$ = 0.2

2. It means a 5 in 100 chance of an event happening (the same as 1 in 20).

3. a. 1%   **b.** 10%
   c. 5%   **d.** 0.1%
   e. 3%

4. a. $\dfrac{4}{52}$ or $\dfrac{1}{13}$

b. 3 cards × 4 suits divided by the total number of cards = $\dfrac{12}{52}$ = 0.23

c. $\dfrac{2}{52}$ or $\dfrac{1}{26}$ = 0.04

3.2 Hypothesis testing

1. non-directional

2. The hypothesis states the direction the results go in (the alcohol decreases reaction time).

3. a. Directional because the researcher has stated the direction the results will go in (being watched will result in people solving the task more slowly).

b. There is no difference in how people solve difficult problems when they are observed by others compared to when they are alone.

4. a. It is not a null hypothesis because the researcher is stating that there is a difference.

b. There is no difference in the number of anagrams solved in silence compared to when music is playing.

3.3 Error types

1. 0.01 is a stringent significance level used to make sure there is very little chance of reaching the wrong conclusion (a Type 1 error). This means you may accept a null hypothesis that is in fact false (a Type 2 error). A level of 0.05 is less stringent and so a Type 2 error is less likely.

2. It is a good compromise between being too stringent or lenient, and means you more often avoid Type 1 and Type 2 errors.

3. a. There is a 5% chance that the null hypothesis will be mistakenly rejected.

b. There would have been a 1% chance that the null hypothesis would be mistakenly rejected.

c. You are more likely to falsely accept the null hypothesis (Type 2 error) at a stringent significance level of 1%.

3.4 Using statistical tests

1. Statements about the population can only be made using inferential statistics.

2. *rho*

3. a. Hypothesis is directional or non-directional/one- or two-tailed test;

b. the desired significance level;

c. the degrees of freedom or N value;

d. calculated/observed value;

e. critical value;

f. whether the calculated value needs to be greater or less than critical value.

4. a. A one-tailed test is required. With a significance level of 0.01, the critical value for 8 participants is 1 – the calculated value is greater so the null hypothesis must be accepted.

 b. A two-tailed test is required with the significance level of 0.1, the critical value for 20 participants is 60 – the calculated value is smaller so the null hypothesis can be rejected at the 10% (and 5%) significance level.

3.5 The sign test

1. 0.01

2. *df* will be 17 because there is one zero score. Therefore, the critical value for *S* is 4. The calculated value is not less than or equal to the critical value, therefore the result is not significant.

3. a. There is no difference in the colours selected by twin pairs.

 b. From left to right: −, +, +, +, −, +, −, −, −

 S is the smaller value = 4

 c. Two-tailed hypothesis with *N* = 8, the calculated value for *S* = 4, and at the 5% level the critical value = 0

 As the calculated value for *S* is larger than the critical value at the 5% level, you must accept the null hypothesis (p ≤ 0.05) and conclude that there is no difference in the colours selected by twin pairs.

3.6 Wilcoxon test and
3.7 Mann–Whitney test

1. a. Related because the same person was asked twice.

 b. It is ordinal because the students have been asked to rate their liking and the difference between each rating is not equal.

 c. Wilcoxon test should be used because a test of difference is required (data sets are being compared for differences in ratings after one month and six months), the two samples of data are related, and the data are ordinal (they are ratings and the difference between each rating is not equal).

d.

Rating	Frequency	Ranks	Final rank
2	I	1	1
3	II	2, 3	2.5
4	ЖЖ	4, 5, 6, 7, 8	6
5	ЖЖ III	9, 10, 11, 12, 13, 14, 15, 16	12.5
6	ЖЖ II	17, 18, 19, 20, 21, 22, 23	20
7	ЖЖ	24, 25, 26, 27, 28	26
8	ЖЖ I	29, 30, 31, 32, 33, 34	31.5
9	IIII	35, 36, 37, 38	36.5
10	II	39, 40	39.5

2. a. The data could be classed as interval because the scores on the test are equal. However, it could be argued that they are not equal because some test items are more difficult than others, for example, a decrease from 15 to 13 is not the same as a decrease from 18 to 16. In this case the level of measurement should be classed as ordinal.

 b. Wilcoxon test as a test of difference is required (data sets are being compared for differences before and after treatment), the two samples of data are related, and the data are ordinal or better (see part a).

 c. There is no difference in depression score before and after treatment.

d.

Score before	Score after	Difference	Rank
15	16	−1	3
12	15	−3	6
12	11	1	3
19	13	6	7
15	14	1	3
17	10	7	8
10	11	−1	3
18	19	−1	3
16	8	8	9
15	15	0	

Step 1: Decide which test to use. See answer to part b.

Step 2: State the hypotheses. See question and answer to part c.

Step 3: Record the data and calculate differences and ranks. See table on previous page.

Step 4: Find calculated value of T.
Add up the ranks for positive differences:
$3 + 7 + 3 + 8 + 9 = 30$
Add up the ranks for negative differences:
$3 + 6 + 3 + 3 = 15$
T is the smaller value $= 15$

Step 5: Is the result in the right direction? The predicted direction was that scores after treatment would be lower. The ranks are higher for positive differences, which means most are in the right direction (score on depression test decreased showing improvement).

Step 6: Find critical value of T.

- One-tailed or two-tailed test? The hypothesis is directional, therefore a one-tailed test is used.

- Significance level selected $= p \leq 0.05$ (5% level)

- Degrees of freedom (df) or N value (total number of scores ignoring zero values) $= N = 9$ (1 score omitted)

  Now you use the table of critical values (on page 63), locate the column headed 0.05 for a one-tailed test and the row which begins with $N = 9$

- The critical value of $T = 8$

Step 7: Accept or reject the null hypothesis and draw a conclusion.

For the Wilcoxon test the calculated (observed) value must be equal to or less than the critical value for the result to be significant.

The calculated value (15) is not equal to or less than the critical value (8) so the result is not significant at $p \leq 0.05$ (5% significance level).

This means you must accept the null hypothesis and conclude that there is no difference in depression score before and after treatment. In other words, the treatment is not effective ($p \leq 0.05$).

3. a. A Mann-Whitney test as a test of difference is required (data sets are being compared for differences in memory recall), the two samples of data are unrelated (one set is younger participants and the other set is older participants), and the data are ordinal or better (items on a memory test could be argued to vary in difficulty so might be classed as ordinal).

b. non-directional

c. There is no difference in the accuracy scores of younger and older participants on a task assessing the accuracy of memory.

d.

Frequency table to calculate ranks

Score	Frequency	Ranks	Final rank
1	II	1, 2	1.5
2	II	3, 4	3.5
3	III	5, 6, 7	6
4	IIII	8, 9, 10, 11	9.5
5	II	12, 13	12.5
6	II	14, 15	14.5
7	III	16, 17, 18	17
8	II	19, 20	19.5
9	II	21, 22	21.5
10	I	23	23

Final data table

Condition A Younger participants	Rank A	Condition B Older participants	Rank B
3	6	6	14.5
10	23	3	6
5	12.5	7	17
7	17	5	12.5
4	9.5	2	3.5
9	21.5	4	9.5
1	1.5	8	19.5
6	14.5	9	21.5
2	3.5	7	17
3	6	8	19.5
4	9.5	4	9.5
		1	1.5
$N_A = 11$ $\Sigma R_A = 124.5$		$N_B = 12$ $\Sigma R_B = 151.5$	

Step 1: Decide which test to use.

See answer to part a.

Step 2: State the hypotheses.

See question and answer to part c.

Step 3: Record the data and calculate differences and ranks. See tables on previous page.

Step 4: Find calculated value of U.

You substitute your values in the formula:

$$U_A = 11 \times 12 + \frac{11(11+1)}{2} - 124.5$$

$$= 132 + 66 - 124.5 = 73.5$$

$$U_B = 11 \times 12 + \frac{12(12+1)}{2} - 151.5$$

$$= 132 + 78 - 151.5 = 58.5$$

U = the smaller of the two values = 58.5

Step 5: Is the result in the right direction? There is no stated direction.

Step 6: Find critical value of U.

- One-tailed or two-tailed test? The hypothesis is non-directional, therefore a two-tailed test is used.
- Significance level selected = $p \le 0.05$ (5% level)
- Degrees of freedom (df) or N value = $N_A = 11$ and $N_B = 12$

Now you use the table of critical values for a two-tailed test (see page 70) and locate the row and column that begin with your N values.

- The critical value of $U = 33$

Step 7: Accept or reject the null hypothesis and draw a conclusion.

For the Mann–Whitney test the calculated (observed) value must be equal to or less than the critical value for your result to be significant.

The calculated value (58.5) is not equal to or less than the critical value (33) so your result is not significant at $p \le 0.05$ (5% significance level).

This means you must accept the null hypothesis and conclude that there is no difference in the accuracy scores of younger and older participants on a task assessing the accuracy of memory ($p \le 0.05$).

4. **a.** Mann–Whitney
   b. Mann–Whitney
   c. Wilcoxon
   d. Mann–Whitney

5. **a.** Accept the null hypothesis (critical value = 25).
   b. Accept the null hypothesis (critical value = 3).
   c. Reject the null hypothesis (critical value = 53).
   d. Reject the null hypothesis (critical value = 15).

6. **a.** Reject the null hypothesis (critical value = 11).
   b. Accept the null hypothesis (critical value = 46).
   c. Reject the null hypothesis (critical value = 17).
   d. Reject the null hypothesis (critical value = 44).

3.8 Related t-test and 3.9 Unrelated t-test

1. **a.** Unrelated because there are two separate groups of participants.
   b. It is classed as interval because the test questions could be assumed to be of equal difficulty.
   c. Unrelated t-test because a test of difference is required (data sets are being compared for differences in memory recall), the two samples of data are unrelated (two separate groups of participants tested), and the data are interval (test questions would be assumed to be of equal difficulty).
   d. One-tailed because the hypothesis is directional.
   e. $df = 16 + 11 - 2 = 25$, so critical value is 1.708

2. Related t-test because a test of difference is required (data sets are being compared for differences in memory recall), the two samples of data are related (each participant is tested twice), and the data are interval (test questions could be assumed to be of equal difficulty).

3. **a.** Non-directional because the researchers were not sure which direction the results would be in.
   b. $t = \dfrac{8}{\sqrt{\left(\dfrac{(10 \times 66 - 8 \times 8)}{9}\right)}}$

   $= \dfrac{8}{8.138} = 0.983$
   c. The critical value for a non-directional hypothesis (two-tailed test) at $p = 0.02$ with 10 participants is 2.764 – the calculated value is not equal to or greater than this so the null hypothesis must be accepted.

4. a. Mann–Whitney, because a test of difference is required (as data sets are being compared for differences in IQ scores), the two samples of data are unrelated (older versus younger children), and the data are ordinal rather than interval (IQ tests use standardised scores and not equal intervals).

b. Sign test, because a test of difference is required (as two data items are being compared for differences in colour blindness), the two samples of data are related (twins are matched participants), and the data are nominal (colour blind, yes or no).

c. Wilcoxon (or the sign test), because a test of difference is required (as data sets are being compared for differences in ratings), the two samples of data are related (from the same person), and the data are ordinal (rating scores do not represent equal differences between ratings) or the data could be classed as categorical (food more attractive or food not more attractive after eating).

d. Unrelated *t*-test (or Mann–Whitney), because a test of difference is required (as data sets are being compared for differences in memory test performance), the two samples of data are unrelated (two groups of participants), and the data are interval (test items can be assumed to be of equal difficulty) or assume data are ordinal (test items not equal in difficulty).

e. Mann–Whitney because a test of difference is required (as data sets are being compared for differences in aggressiveness), the two samples of data are unrelated (either children whose mothers work or children whose mothers don't work), and the data are ordinal (rating scores where the intervals may not be equivalent).

f. Related *t*-test because a test of difference is required (as data sets are being compared for differences in heart rate), the two samples of data are related (same person compared), and the data are interval (heart rate is measured on a scale with equal intervals).

5. For a related *t*-test *df* equals the number of participants minus any pairs where there was a zero difference. For an unrelated *t*-test *df* equals the sum of participants in the two separate groups less 2.

6. No, even though a significant result was found they must accept the null hypothesis because the alternative hypothesis was not supported.

7. a. Critical value = 1.833, therefore reject null hypothesis (1.870 is significant).

b. Critical value = 1.746, therefore reject null hypothesis (1.93 is significant).

c. Critical value = 2.583, therefore accept null hypothesis (1.93 is not significant).

d. Critical value = 1.761, therefore accept null hypothesis (1.33 is not significant).

3.10 Chi-squared test and 3.11 Spearman's test

1. The data items are not independent as each person voted twice.

2. It is not significant at the 2% level (critical value = 5.41), but is significant at the 5% level (critical value = 3.84).

3. a. For example, 'Age is associated with whether a person supports Labour or Conservatives' or the alternative hypothesis could be 'There is a difference between young and old voters in terms of support for Labour or the Conservatives'.

b.

	Labour	Conservative	Total
Under 30			
Over 30			
Total			

c. $(2 - 1) \times (2 - 1) = 1$

d. Equal to or greater than 2.706 for a two-tailed test.

4. a. You would need 2 rows (one for girls and one for boys) and 3 columns (for the 3 psychologists). The first number represents the number of rows.

b.

	Milgram	Bandura	Baddeley	Total
Girls	18	5	8	31
Boys	12	6	1	19
Total	30	11	9	50

5. a. There is an association between parenting style (authoritarian and permissive) and a person's obedience level (high or low).

b. There is no association between parenting style (authoritarian and permissive) and a person's obedience level (high or low).

c. The hypothesis states an association between two variables (parenting style and obedience), the data in each cell are independent, and the data are nominal because each person is classified in terms of one category.

d.

	Row total × column total / total = expected frequency (E)	Subtract expected value (E) from observed frequency (O) (O – E)	Square previous value (O – E)²	Divide previous value by expected frequency (O – E)² / E
Cell A	$42 \times \dfrac{38}{75} = 21.28$	$23 - 21.28 = 1.72$	2.9584	0.1390
Cell B	$42 \times \dfrac{37}{75} = 20.72$	$19 - 20.72 = -1.72$	2.9584	0.1428
Cell C	$33 \times \dfrac{38}{75} = 16.72$	$15 - 16.72 = -1.72$	2.9584	0.1769
Cell D	$33 \times \dfrac{37}{75} = 16.28$	$18 - 16.28 = 1.72$	2.9584	0.1817
				$\Sigma (O-E)2/E = 0.6404$

e. That it is a two-tailed test, the significance level is 5%, and degrees of freedom = 1.

f. 3.841

g. The null hypothesis should be accepted because the observed value is less than the critical value.

6. a. Spearman's test because you are looking for a correlation between two sets of data, and the items are paired because they are from couples. The data count as ordinal because a rating scale is used and you cannot assume there are equal intervals between the different ratings.

b. The ratings for each couple are positively correlated (directional). (You would not expect a negative correlation because that would mean that the more attractive a person is, the less attractive their partner would be.)

c.

Couple	Rating for man	Rank	Rating for woman	Rank	d	d²
A	9	9	8	8	1	1
B	5	3.5	7	6	−2.5	6.25
C	5	3.5	6	3.5	0	0
D	4	2	7	6	−4	16
E	8	7.5	9	9	−1.5	2.25
F	7	6	6	3.5	2.5	6.25
G	8	7.5	7	6	1.5	2.25
H	3	1	2	1	0	0
I	6	5	4	2	3	9
N = 9						Σd² = 43

$$rho = 1 - \frac{(6 \times 43)}{(9 \times (9 \times 9 - 1))}$$

$$= 1 - \frac{258}{720} = 0.642 \text{ (3 d.p.)}$$

d. You can reject the null hypothesis because the critical value for a one-tailed test (directional hypothesis) at 5% level and N = 9 is .600. Your calculated value is greater than the critical value.

e. It would mean that there was a negative correlation between couples – more attractive people were coupled with less attractive people. It also means that you would have to accept the null hypothesis even if the calculated value was greater than or equal to the critical value because the results are in the wrong direction.

f. To decide this you need to consult the critical values table for a two-tailed test at N = 9 and a 5% significance level. The critical value is .700 – the calculated value is less than this so the test is not significant and you must accept the null hypothesis.

g. You could sketch a scatter diagram.

7. a. You are analysing a correlation, the data are related, and the IQ data are classed as ordinal because IQ is a standardised score.

b. $rho = 1 - \dfrac{(6 \times 31)}{(8 \times (8 \times 8 - 1))}$

$= 1 - \dfrac{186}{504} = +0.631$ (3 d.p.)

c. For a one-tailed test (a positive correlation is expected) at a 5% level of significance and $N = 8$ the critical value is .643. As the calculated value is less than this, the result is not significant.

3.12 Pearson's test

1. Parametric tests have greater power which means they are more likely to find significance that might be missed when using a non-parametric test.

2. You are looking at a correlational analysis, the data are related, and are interval because they use a scale of measurement with equal intervals.

3. **a.** $df = 16$, the critical value is 0.400, the calculated value is greater than the critical value and therefore the null hypothesis is rejected.

   b. $df = 4$, the critical value is 0.729, the calculated value is less than the critical value and therefore the null hypothesis is accepted.

   c. $df = 9$, the critical value is 0.602, the calculated value is less than the critical value and therefore the null hypothesis is accepted.

Summary questions for Section 3

1. **a.** Wilcoxon or sign test, because (two points from) the data are at least at the ordinal level of measurement, a test of difference is required because two sets of scores are being compared to see which is larger, the scores are related because the experimental design is matched pairs.

   b. A directional hypothesis specifies the direction of a difference or correlation, whereas a non-directional hypothesis only states that here will be a difference or correlation.

   c. There is a probability of 0.05 or less (5% chance or less) of seeing results as extreme as this if the null hypothesis was true.

2. **a.** Directional, because the researcher is stating the expected direction of the correlation (positive).

   b. Spearman's *rho* because the researcher is looking for a correlation and the data are at least at the ordinal level of measurement (the IQ scores are formed from test items that are not equivalent).

   c. There is a 5% chance that the null hypothesis was mistakenly rejected (i.e., a Type 1 error).

3. **a.** (Two from) an independent groups design was used, the test looked for a difference between the groups, and the data were at least at the ordinal level of measurement (some of the words might have been easier to remember than others).

   b. The researcher was looking for a difference between groups. Spearman's *rho* is used for studies of correlation, which this study is not looking at.

   c. When $N_A = 7$ and $N_B = 6$, the critical value for a two tailed test at the 5% significance level is 6. The calculated value (3) must be equal to, or less than, the critical value for significance. As 3 is less than 6, so the result is significant.

4. **a.** directional

   b. (Two from) the data could be classed as nominal (Taylor Swift, Pink Floyd), repeated measures are used, and the teacher is looking for a difference between groups.

   c. $S = 5$. This is determined by converting the differences or outcomes to signs (+ or −) and selecting the least frequently occurring sign (ignoring differences of zero).

   d. The calculated value is significant. With $N = 20$ and two zero values, $df = 18$, for a one-tailed test at the 0.05 level of significance the critical value is 5. The calculated value of S (5) must be equal to or less than the critical value (5). The calculated value is equal to the critical value, so the difference in groups is significant.

5. **a.** It is nominal because frequency data are recorded.

   b. χ^2 (or sign test)

   c. There is a 1% chance that the null hypothesis would have been mistakenly rejected (Type 1 error).

Index

A
algebraic equations   **22–3**
alternative (alternate) hypothesis   **55, 61**
arithmetic
    decimals and decimal places   **8–9**
    estimating results   **16–17**
    fractions   **6–7**
    percentages   **10–11**
    ratios   **12–13**
    standard form   **14–15**

B
bar charts   **43–5**
    interpreting a bar chart   **44**

C
calculated values   **62, 63, 66, 68, 73, 75, 79, 82, 85**
    rule of R   **74**
calculators   **8, 38**
chi-squared test   **78–80**
common factors   **6**
conclusions   **62, 64, 66, 70, 74, 76, 80, 83, 86**
contingency tables   **79**
correlations   **50–2**
    correlation coefficients   **81**
    curvilinear relationship   **51**
    moderate negative correlation   **51**
    moderate positive correlation   **51**
    no correlation   **51**
    strong negative correlation   **51**
    strong positive correlation   **51**
critical values   **62, 64, 66, 68–9, 73–4, 75, 79–80, 82, 85–6**
curvilinear relationship   **51**

D
decimals   **8**
    changing a decimal fraction to a percentage   **10**
    decimal places   **8, 20**
    giving your answer to a set number of decimal places   **9**
    moving the decimal point to divide or multiply   **8**
    working out the decimal equivalent of a fraction   **8**
    working out the decimal fraction of a whole   **8**
    working out the fraction equivalent of a decimal   **8**
denominators   **6, 10**
descriptive statistics   **33–34, 37–38, 41–52, 61**
directional hypotheses   **57–8**
distribution   **46–49**
    normal distribution   **46–7**
    skewed distributions   **48–9**

E
error types   **59–60**
    Type 1 and Type 2 errors   **59**
estimating results   **16–17**
event sampling   **31–2**
exponents   **14, 15**

F
fractions   **6, 8**
    changing a decimal fraction to a percentage   **10**
    changing a fraction to a percentage   **10**
    changing a percentage to a fraction   **10**
    simplifying a fraction   **6**
    working out a fraction   **6**
    working out more complex fractions   **6–7**
    working out the decimal equivalent of a fraction   **8**
    working out the decimal fraction of a whole   **8**
    working out the fraction equivalent of a decimal   **8**
frequency diagrams   **44–48**
    constructing a frequency diagram   **44**
frequency tables   **41**

G
graphs   **44–52**

H
histograms   **43–5**
    interpreting a histogram   **44**
hypothesis testing   **57–8**

I
independent data   **67, 75**
inferential tests
    chi-squared test   **78–80**
    error types   **59–60**
    hypothesis testing   **57–8**
    Mann-Whitney test   **67–71**
    Pearson's test   **85–6**
    related t-test   **72–4**

sign test **63–4**
simple probability and the null hypothesis **55–6**
Spearman's test **81–4**
unrelated *t*-test **75–7**
using inferential tests **61–2**
    accept or reject the null hypothesis and draw
      a conclusion **62**
    decide which test to use **61**
    find the calculated value **62**
    find the critical value **62**
    is the result in the right direction? **62**
    record data and preliminary calculations **61**
    state the hypotheses (alternative and null) **61**
Wilcoxon test **65–6**
inferential statistical tests *see Inferential tests*
interval measurement **35**

L
levels of measurement **35–6**

M
Mann-Whitney test **22, 67–71**
    accept or reject the null hypothesis and draw a
      conclusion **70**
    find calculated value of *U* **68**
    find critical value of *U* **68**
    is the result in the right direction? **68**
    reasons for choosing the Mann-Whitney
      test **67**
    record the data and calculate differences
      and ranks **67–8**
    state the hypotheses **67**
    table of critical values for a one-tailed test **69**
    table of critical values for a two-tailed test **69**
mantissas **14**
mathematical symbols **21**
mean **33**
measures of central tendency **33–4**
measures of dispersion **33, 37–8**
median **33, 34**
mode (modal group) **33, 34**

N
negative skew **48–9**
nominal measurement **35**
non-directional hypotheses **57–8**
non-parametric tests **65**
normal distribution **46–7**
null hypothesis **55, 61, 62, 64, 66, 70, 74, 76, 80,
    83, 86**
numerators **6, 10**

O
observed values **62, 74**
opportunity samples **328**
order of magnitude calculations **18**
    comparing big numbers **18**
    comparing small numbers **18**
ordinal measurement **35**

P
paired data **63, 65, 72**
parametric tests **65**
    parametric data criteria **72**
part-to-part ratios **12**
part-to-whole ratios **12**
Pearson's test **85–6**
    accept or reject the null hypothesis and draw
      a conclusion **86**
    find the calculated value of *r* **85**
    find the critical value of *r* **85**
    is the result in the right direction? **85**
    reasons for choosing Pearson's test **85**
    record the data **85**
    state the hypotheses **85**
    table of critical values for Pearson's *r* **86**
percentages **10**
    changing a decimal fraction to a percentage **10**
    changing a fraction to a percentage **10**
    changing a percentage to a fraction **10**
    using a percentage to calculate an answer **11**
    working out a percentage **10**
positive skew **48–9**
primary data **27–8**
probability **55**
    level of probability **56**
    why probability matters in psychology **56**

Q
quantitative and qualitative data **25–6**
quota samples **7, 13, 30**

R
random samples **29**
range **37, 38**
ranking data **42, 66, 67–8, 81**
ratios **12**
    expressing a part-to-part relationship **12**
    expressing a part-to-whole relationship **12**
    ratio measurement **35**
    using ratios in calculations **13**
related data **63, 65, 72**

related *t*-test **22, 72–4**
    accept or reject the null hypothesis and draw
        a conclusion **74**
    find critical value of *t* **73**
    find the calculated value of *t* **73**
    is the result in the right direction? **73**
    parametric data criteria **72**
    reasons for choosing a related *t*-test **72**
    record data and calculate difference **72–3**
    state the hypotheses
    table of critical values **74**
rounding down **9**
rounding up **9**
rule of R **74**

S
sampling observations **31–2**
sampling participants **29**
    opportunity samples **328**
    random samples **29**
    self-selecting (volunteer) samples **29**
    snowball samples **29–30**
    stratified and quota samples **30**
    systematic samples **30**
scatter diagrams **50–2**
secondary data **27–8**
self-selecting samples **29**
sign test **63–4**
    accept or reject the null hypothesis and draw
        a conclusion **64**
    find the calculated value of *S* **63**
    find the critical value of *S* **64**
    is the result in the right direction? **63**
    reasons for choosing the sign test **63**
    record data and signs **63**
    state the hypotheses **63**
significance levels **59, 62**
significant figures **19–20**
skewed distributions **48–9**
snowball samples **29–30**
Spearman's test **22, 81–4**
    accept or reject the null hypothesis and draw
        a conclusion **83**
    find the calculated value of *rho* **82**
    find the critical value of *rho* **82**
    is the result in the right direction? **82**
    reasons for choosing Spearman's
        test **81**
    record the data and calculate ranks and
        differences **81**
    state the hypotheses **81**

table of critical values for Spearman's
    rho **82**
standard deviation **37, 38**
    interpreting the standard deviation **38**
standard form **14**
    changing a very large number to standard
        form **14**
    changing a very small number to standard
        form **15**
    changing standard form back to a very large
        number **15**
statistical tests *see Inferential tests*
stratified samples **7, 13, 30**
strong negative correlation **51**
strong positive correlation **51**
systematic samples **30**

T
tables **33, 41**
time sampling **32**
Type 1 errors **60–61**
Type 2 errors **60–61**

U
unrelated data **67, 75**
unrelated *t*-test **75–7**
    accept or reject the null hypothesis and draw
        a conclusion **76**
    find the calculated value of *t* **75**
    find the critical value of *t* **75**
    is the result in the right direction? **75**
    reasons for choosing unrelated *t*-test **75**
    record the data **75**
    state the hypotheses **75**
V
volunteer samples **29**
W
whole numbers **8**
Wilcoxon test **65–6, 67, 74**
    accept or reject the null hypothesis and draw
        a conclusion **66**
    find the calculated value of *T* **66**
    find the critical value of *T* **66**
    is the result in the right direction? **66**
    non-parametric tests **65**
    reasons for choosing the Wilcoxon test **65**
    record the data and calculate the differences
        and ranks **65–6**
    state the hypotheses **65**
    table of critical values **62**

Notes

OXFORD

UNIVERSITY PRESS

Great Clarendon Street, Oxford, OX2 6DP, United Kingdom

Oxford University Press is a department of the University of Oxford.
It furthers the University's objective of excellence in research, scholarship,
and education by publishing worldwide. Oxford is a registered trade mark of
Oxford University Press in the UK and in certain other countries

British Library Cataloguing in Publication Data
Data available

978-0-19-843790-1

10 9 8 7 6 5 4 3 2 1

Printed and bound by CPI Group (UK) Ltd, Croydon, CR0 4YY

Acknowledgements

Cover: Shutterstock/Gular Samadova; Shutterstock/grebeshkovmaxim

Artwork by Thomson Digital

The author would like to thank Sarah Flynn for all her support with this project.

The publisher would like to thank Clare Compton for contributions
to this work.

Although we have made every effort to trace and contact all
copyright holders before publication this has not been possible in all
cases. If notified, the publisher will rectify any errors or omissions at
the earliest opportunity.

Links to third party websites are provided by Oxford in good faith
and for information only. Oxford disclaims any responsibility for
the materials contained in any third party website referenced in
this work.